THE PHILOSOPHY OF MATHEMATICS

THE SCIENCE OF MATHEMATICS.

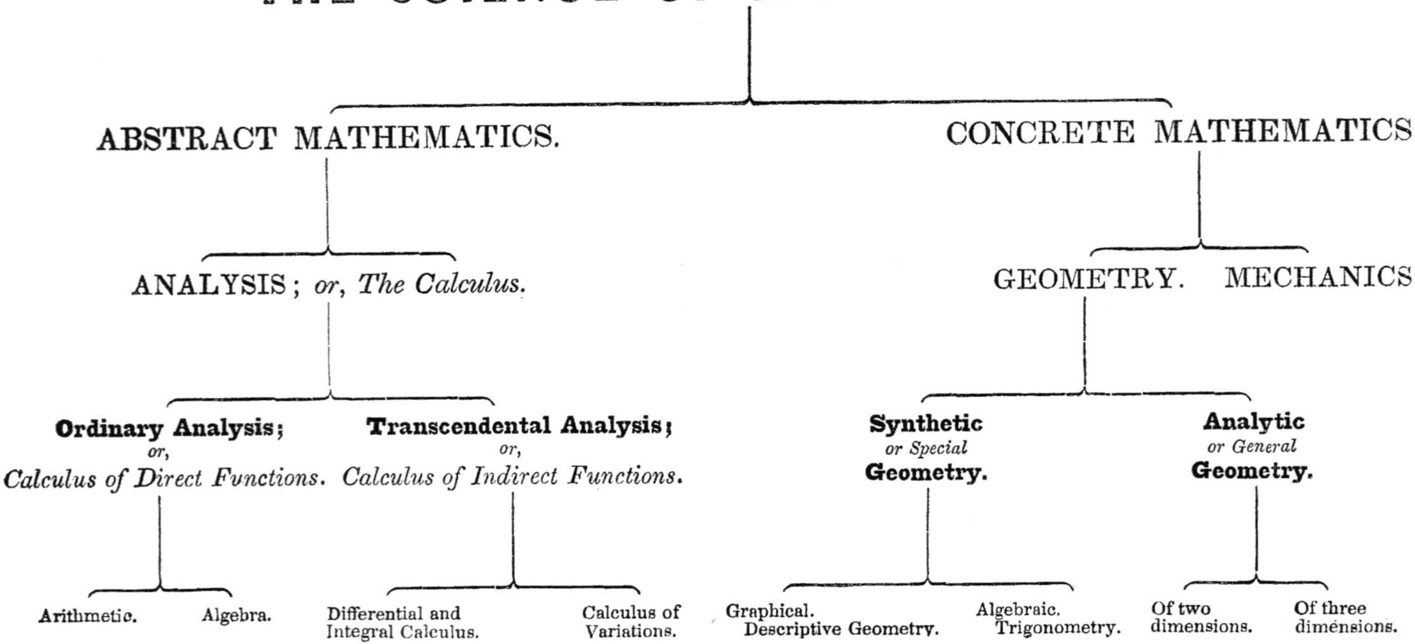

The Philosophy of MATHEMATICS

Translated from the
Cours de Philosophie Positive
of

AUGUSTE COMTE

by

W. M. GILLESPIE

Professor of Civil Engineering & Adj. Prof.
of Mathematics in Union College

DOVER PUBLICATIONS, INC.
Mineola, New York

DOVER PHOENIX EDITIONS

Bibliographical Note

This Dover edition, first published in 2005, is an unabridged republication of the work first published in 1851 by Harper & Brothers, New York. The original French edition of the complete *Cours de Philosophie Positive* was published by Bachelier, Paris. 1830–42.

International Standard Book Number: 0-486-44249-7

Manufactured in the United States of America
Dover Publications, Inc., 31 East 2nd Street, Mineola, N.Y. 11501

PREFACE.

THE pleasure and profit which the translator has received from the great work here presented, have induced him to lay it before his fellow-teachers and students of Mathematics in a more accessible form than that in which it has hitherto appeared. The want of a comprehensive map of the wide region of mathematical science—a bird's-eye view of its leading features, and of the true bearings and relations of all its parts—is felt by every thoughtful student. He is like the visitor to a great city, who gets no just idea of its extent and situation till he has seen it from some commanding eminence. To have a panoramic view of the whole district—presenting at one glance all the parts in due co-ordination, and the darkest nooks clearly shown—is invaluable to either traveller or student. It is this which has been most perfectly accomplished for mathematical science by the author whose work is here presented.

Clearness and depth, comprehensiveness and precision, have never, perhaps, been so remarkably united as in AUGUSTE COMTE. He views his subject from an elevation which gives to each part of the complex whole its true position and value, while his telescopic glance loses none of the needful details, and not only itself pierces to the heart of the matter, but converts its opaqueness into such transparent crystal, that other eyes are enabled to see as deeply into it as his own.

Any mathematician who peruses this volume will need no other justification of the high opinion here expressed; but others may appreciate the following endorsements of well-known authorities. *Mill*, in his "Logic," calls the work of M. Comte "by far the greatest yet produced on the Philosophy of the sciences;" and adds, "of this admirable work, one of the most admirable portions is that in which he may truly be said to have created the Philosophy of the higher Mathematics:" *Morell*, in his "Speculative Philosophy of Europe," says, "The classification given of the sciences at large, and their regular order of development, is unquestionably a master-piece of scientific thinking, as simple as it is comprehensive;" and *Lewes*, in his "Biographical History of Philosophy," names Comte "the Bacon of the nineteenth century," and says, "I unhesitatingly record my conviction that this is the greatest work of our age."

The complete work of M. Comte—his "*Cours de Philosophie Positive*"—fills six large octavo volumes, of six or seven hundred pages each, two thirds of the first volume comprising the purely mathematical portion. The great bulk of the "Course" is the probable cause of the fewness of those to whom even this section of it is known. Its presentation in its present form is therefore felt by the translator to be a most useful contribution to mathematical progress in this country.

The comprehensiveness of the style of the author—grasping all possible forms of an idea in one Briarean sentence, armed at all points against leaving any opening for mistake or forgetfulness—occasionally verges upon cumbersomeness and formality. The translator has, therefore, sometimes taken the liberty of breaking up or condensing a long sentence, and omitting a few passages not absolutely necessary, or referring to the peculiar "Positive philosophy" of the author; but he has generally aimed at a conscientious fidelity to the original. It has often been difficult to retain its fine shades and subtile distinctions of meaning, and, at the same time, replace the peculiarly appropriate French idioms by corresponding English ones. The attempt, however, has always been made, though, when the best course has been at all doubtful, the language of the original has been followed as closely as possible, and, when necessary, smoothness and grace have been unhesitatingly sacrificed to the higher attributes of clearness and precision.

Some forms of expression may strike the reader as unusual, but they have been retained because they were characteristic, not of the mere language of the original, but of its spirit. When a great thinker has clothed his conceptions in phrases which are singular even in his own tongue, he who professes to translate him is bound faithfully to preserve such forms of speech, as far as is practicable; and this has been here done with respect to such peculiarities of expression as belong to the author, not as a foreigner, but as an individual—not because he writes in French, but because he is Auguste Comte.

The young student of Mathematics should not attempt to read the whole of this volume at once, but should peruse each portion of it in connexion with the temporary subject of his special study: the first chapter of the first book, for example, while he is studying Algebra; the first chapter of the second book, when he has made some progress in Geometry; and so with the rest. Passages which are obscure at the first reading will brighten up at the second; and as his own studies cover a larger portion of the field of Mathematics, he will see more and more clearly their relations to one another, and to those which he is next to take up. For this end he is urgently recommended to obtain a perfect familiarity with the "Analytical Table of Contents," which maps out the whole subject, the grand divisions of which are also indicated in the Tabular View facing the title-page. Corresponding heads will be found in the body of the work, the principal divisions being in SMALL CAPITALS, and the subdivisions in *Italics*. For these details the translator alone is responsible.

ANALYTICAL TABLE OF CONTENTS.

INTRODUCTION.

BOOK I.

ANALYSIS.

CHAPTER I.

CHAPTER II.

CHAPTER III.

TRANSCENDENTAL ANALYSIS:

CHAPTER IV.

CHAPTER V.

CHAPTER VI.

BOOK II.

GEOMETRY.

CHAPTER I.

CHAPTER II.

ANCIENT OR SYNTHETIC GEOMETRY

CHAPTER III.

MODERN OR ANALYTICAL GEOMETRY

THE

PHILOSOPHY OF MATHEMATICS.

INTRODUCTION.

GENERAL CONSIDERATIONS.

Although Mathematical Science is the most ancient and the most perfect of all, yet the general idea which we ought to form of it has not yet been clearly determined. Its definition and its principal divisions have remained till now vague and uncertain. Indeed the plural name—"The Mathematics"—by which we commonly designate it, would alone suffice to indicate the want of unity in the common conception of it.

In truth, it was not till the commencement of the last century that the different fundamental conceptions which constitute this great science were each of them sufficiently developed to permit the true spirit of the whole to manifest itself with clearness. Since that epoch the attention of geometers has been too exclusively absorbed by the special perfecting of the different branches, and by the application which they have made of them to the most important laws of the universe, to allow them to give due attention to the general system of the science.

But at the present time the progress of the special departments is no longer so rapid as to forbid the contemplation of the whole. The science of mathematics is now sufficiently developed, both in itself and as to its most essential application, to have arrived at that state of consistency in which we ought to strive to arrange its different parts in a single system, in order to prepare for new advances. We may even observe that the last important improvements of the science have directly paved the way for this important philosophical operation, by impressing on its principal parts a character of unity which did not previously exist.

To form a just idea of the object of mathematical science, we may start from the indefinite and meaningless definition of it usually given, in calling it "*The science of magnitudes*," or, which is more definite, "*The science which has for its object the measurement of magnitudes.*" Let us see how we can rise from this rough sketch (which is singularly deficient in precision and depth, though, at bottom, just) to a veritable definition, worthy of the importance, the extent, and the difficulty of the science.

THE OBJECT OF MATHEMATICS.

Measuring Magnitudes. The question of *measuring* a magnitude in itself presents to the mind no other idea than that of the simple direct comparison of this magnitude with another similar magnitude, supposed to be known, which it takes for the *unit* of comparison among all others of the same kind. According to this definition, then, the science of mathematics—vast and profound as it is with reason reputed to be—instead of being an immense concatenation of prolonged mental labours, which offer inexhaustible occupation to our intellectual activity, would seem to consist of a simple series of mechanical processes for obtaining directly the ratios of the quantities to be measured to those by which we wish to measure them, by the aid of operations of similar character to the superposition of lines, as practiced by the carpenter with his rule.

The error of this definition consists in presenting as direct an object which is almost always, on the contrary, very indirect. The *direct* measurement of a magnitude, by superposition or any similar process, is most frequently an operation quite impossible for us to perform; so that if we had no other means for determining magnitudes than direct comparisons, we should be obliged to renounce the knowledge of most of those which interest us.

Difficulties. The force of this general observation will be understood if we limit ourselves to consider specially the particular case which evidently offers the most facility—that of the measurement of one straight line by another. This comparison, which is certainly the most simple which we can conceive, can nevertheless scarcely ever be effected directly. In reflecting on the whole of the conditions necessary to render a line susceptible of a direct measurement, we see that most frequently they cannot be all fulfilled at the same time. The first and the most palpable of these conditions—that of being able to pass over the line from one end of it to the other, in order to apply the unit of measurement to its whole length—evidently excludes at once by far the greater part of the distances which interest us the most; in the first place, all the distances between the celestial bodies, or from any one of them to the earth; and then, too, even the greater number of terrestrial distances, which are so frequently inaccessible. But even if this first condition be found to be fulfilled, it is still farther necessary that the length be neither too great nor too small, which would render a direct measurement equally impossible. The line must also be suitably situated; for let it be one which we could measure with the greatest facility, if it were horizontal, but conceive it to be turned up vertically, and it becomes impossible to measure it.

The difficulties which we have indicated in reference to measuring lines, exist in a very much greater degree in the measurement of surfaces, volumes, velocities, times, forces, &c. It is this fact which makes necessary the formation of mathematical science, as we are going to see; for the human mind has been compelled to renounce, in almost all cases, the direct measurement of magnitudes, and to seek to determine them *indirectly*, and it is thus that it has been led to the creation of mathematics.

General Method. The general method which is constantly employed, and evidently the only one conceivable, to ascertain magnitudes which do not admit of a direct measurement, consists in connecting them with others which are susceptible of being determined immediately, and by means of which we succeed in discovering the first through the relations which subsist between the two. Such is the precise object of mathematical science viewed as a whole. In order to form a sufficiently extended idea of it, we must consider that this indirect determination of magnitudes may be indirect in very different degrees. In a great number of cases, which are often the most important, the magnitudes, by means of which the principal magnitudes sought are to be determined, cannot themselves be measured directly, and must therefore, in their turn, become the subject of a similar question, and so on; so that on many occasions the human mind is obliged to establish a long series of intermediates between the system of unknown magnitudes which are the final objects of its researches, and the system of magnitudes susceptible of direct measurement, by whose means we finally determine the first, with which at first they appear to have no connexion.

Illustrations. Some examples will make clear any thing which may seem too abstract in the preceding generalities.

1. *Falling Bodies.* Let us consider, in the first place, a natural phenomenon, very simple, indeed, but which may nevertheless give rise to a mathematical question, really existing, and susceptible of actual applications—the phenomenon of the vertical fall of heavy bodies.

The mind the most unused to mathematical conceptions, in observing this phenomenon, perceives at once that the two *quantities* which it presents—namely, the *height* from which a body has fallen, and the *time* of its fall—are necessarily connected with each other, since they vary together, and simultaneously remain fixed; or, in the language of geometers, that they are "*functions*" of each other. The phenomenon, considered under this point of view, gives rise then to a mathematical question, which consists in substituting for the direct measurement of one of these two magnitudes, when it is impossible, the measurement of the other. It is thus, for example, that we may determine indirectly the depth of a precipice, by merely measuring the time that a heavy body would occupy in falling to its bottom, and by suitable procedures this inaccessible depth will be known with as much precision as if it was a horizontal line placed in the most favourable circumstances for easy and exact measurement. On other occasions it is the height from which a body has fallen which it will be easy to ascertain, while the time of the fall could not be observed directly; then the same phenomenon would give rise to the inverse question, namely, to determine the time from the height; as, for example, if we wished to ascertain what would be the duration of the vertical fall of a body falling from the moon to the earth.

In this example the mathematical question is very simple, at least when we do not pay attention to the variation in the intensity of gravity, or the resistance of the fluid which the body passes through in its fall. But, to extend the question, we have only to consider the same phenomenon in its greatest generality, in supposing the fall oblique, and in taking into the account all the principal circumstances. Then, instead of offering simply two variable quantities connected with each other by a relation easy to follow, the phenomenon will present a much greater number; namely, the space traversed, whether in a vertical or horizontal direction; the time employed in traversing it; the velocity of the body at each point of its course; even the intensity and the direction of its primitive impulse, which may also be viewed as variables; and finally, in certain cases (to take every thing into the account), the resistance of the medium and the intensity of gravity. All these different quantities will be connected with one another, in such a way that each in its turn may be indirectly determined by means of the others; and this will present as many distinct mathematical questions as there may be co-existing magnitudes in the phenomenon under consideration. Such a very slight change in the physical conditions of a problem may cause (as in the above example) a mathematical research, at first very elementary, to be placed at once in the rank of the most difficult questions, whose complete and rigorous solution surpasses as yet the utmost power of the human intellect.

2. *Inaccessible Distances.* Let us take a second example from geometrical phenomena. Let it be proposed to determine a distance which is not susceptible of direct measurement; it will be generally conceived as making part of a *figure*, or certain system of lines, chosen in such a way that all its other parts may be observed directly; thus, in the case which is most simple, and to which all the others may be finally reduced, the proposed distance will be considered as belonging to a triangle, in which we can determine directly either another side and two angles, or two sides and one angle. Thenceforward, the knowledge of the desired distance, instead of being obtained directly, will be the result of a mathematical calculation, which will consist in deducing it from the observed elements by means of the relation which connects it with them. This calculation will become successively more and more complicated, if the parts which we have supposed to be known cannot themselves be determined (as is most frequently the case) except in an indirect manner, by the aid of new auxiliary systems, the number of which, in great operations of this kind, finally becomes very considerable. The distance being once determined, the knowledge of it will frequently be sufficient for obtaining new quantities, which will become the subject of new mathematical questions. Thus, when

we know at what distance any object is situated, the simple observation of its apparent diameter will evidently permit us to determine indirectly its real dimensions, however inaccessible it may be, and, by a series of analogous investigations, its surface, its volume, even its weight, and a number of other properties, a knowledge of which seemed forbidden to us.

3. *Astronomical Facts.* It is by such calculations that man has been able to ascertain, not only the distances from the planets to the earth, and, consequently, from each other, but their actual magnitude, their true figure, even to the inequalities of their surface; and, what seemed still more completely hidden from us, their respective masses, their mean densities, the principal circumstances of the fall of heavy bodies on the surface of each of them, &c.

By the power of mathematical theories, all these different results, and many others relative to the different classes of mathematical phenomena, have required no other direct measurements than those of a very small number of straight lines, suitably chosen, and of a greater number of angles. We may even say, with perfect truth, so as to indicate in a word the general range of the science, that if we did not fear to multiply calculations unnecessarily, and if we had not, in consequence, to reserve them for the determination of the quantities which could not be measured directly, the determination of all the magnitudes susceptible of precise estimation, which the various orders of phenomena can offer us, could be finally reduced to the direct measurement of a single straight line and of a suitable number of angles.

TRUE DEFINITION OF MATHEMATICS.

We are now able to define mathematical science with precision, by assigning to it as its object the *indirect* measurement of magnitudes, and by saying it constantly proposes *to determine certain magnitudes from others by means of the precise relations existing between them.*

This enunciation, instead of giving the idea of only an *art*, as do all the ordinary definitions, characterizes immediately a true *science*, and shows it at once to be composed of an immense chain of intellectual operations, which may evidently become very complicated, because of the series of intermediate links which it will be necessary to establish between the unknown quantities and those which admit of a direct measurement; of the number of variables coexistent in the proposed question; and of the nature of the relations between all these different magnitudes furnished by the phenomena under consideration. According to such a definition, the spirit of mathematics consists in always regarding all the quantities which any phenomenon can present, as connected and interwoven with one another, with the view of deducing them from one another. Now there is evidently no phenomenon which cannot give rise to considerations of this kind; whence results the naturally indefinite extent and even the rigorous logical universality of mathematical science. We shall seek farther on to circumscribe as exactly as possible its real extension.

The preceding explanations establish clearly the propriety of the name employed to designate the science which we are considering. This denomination, which has taken to-day so definite a meaning by itself signifies simply *science* in general. Such a designation, rigorously exact for the Greeks, who had no other real science, could be retained by the moderns only to indicate the mathematics as *the* science, beyond all others—the science of sciences.

Indeed, every true science has for its object the determination of certain phenomena by means of others, in accordance with the relations which exist between them. Every *science* consists in the co-ordination of facts; if the different observations were entirely isolated, there would be no science. We may even say, in general terms, that *science* is essentially destined to dispense, so far as the different phenomena permit it, with all direct observation, by enabling us to deduce from the smallest possible number of immediate data the greatest possible number of results. Is not this the real use, whether in speculation or in action, of the *laws* which we succeed in discovering among natural phenomena? Mathematical science, in this point of view, merely pushes to the highest possible degree the same kind of researches which are pursued, in degrees more or less inferior, by every real science in its respective sphere.

ITS TWO FUNDAMENTAL DIVISIONS.

We have thus far viewed mathematical science only as a whole, without paying any regard to its divisions. We must now, in order to complete this general view, and to form a just idea of the philosophical character of the science, consider its fundamental division. The secondary divisions will be examined in the following chapters.

This principal division, which we are about to investigate, can be truly rational, and derived from the real nature of the subject, only so far as it spontaneously presents itself to us, in making the exact analysis of a complete mathematical question. We will, therefore, having determined above what is the general object of mathematical labours, now characterize with precision the principal different orders of inquiries, of which they are constantly composed.

Their different Objects. The complete solution of every mathematical question divides itself necessarily into two parts, of natures essentially distinct, and with relations invariably determinate. We have seen that every mathematical inquiry has for its object to determine unknown magnitudes, according to the relations between them and known magnitudes. Now for this object, it is evidently necessary, in the first place, to ascertain with precision the relations which exist between the quantities which we are considering. This first branch of inquiries constitutes that which I call the *concrete* part of the solution. When it is finished, the question changes; it is now reduced to a pure question of numbers, consisting simply in determining unknown numbers, when we know what precise relations connect them with known numbers. This second branch of inquiries is what I call the *abstract* part of the solution. Hence follows the fundamental division of general mathematical science into *two* great sciences—ABSTRACT MATHEMATICS, and CONCRETE MATHEMATICS.

This analysis may be observed in every complete mathematical question, however simple or complicated it may be. A single example will suffice to make it intelligible.

Taking up again the phenomenon of the vertical fall of a heavy body, and considering the simplest case, we see that in order to succeed in determining, by means of one another, the height whence the body has fallen, and the duration of its fall, we must commence by discovering the exact relation of these two quantities, or, to use the language of geometers, the *equation* which exists between them. Before this first research is completed, every attempt to determine numerically the value of one of these two magnitudes from the other would evidently be premature, for it would have no basis. It is not enough to know vaguely that they depend on one another—which every one at once perceives—but it is necessary to determine in what this dependence consists. This inquiry may be very difficult, and in fact, in the present case, constitutes incomparably the greater part of the problem. The true scientific spirit is so modern, that no one, perhaps, before Galileo, had ever remarked the increase of velocity which a body experiences in its fall: a circumstance which excludes the hypothesis, towards which our mind (always involuntarily inclined to suppose in every phenomenon the most simple *functions*, without any other motive than its greater facility in conceiving them) would be naturally led, that the height was proportional to the time. In a word, this first inquiry terminated in the discovery of the law of Galileo.

When this *concrete* part is completed, the inquiry becomes one of quite another nature. Knowing that the spaces passed through by the body in each successive second of its fall increase as the series of odd numbers, we have then a problem purely numerical and *abstract;* to deduce the height from the time, or the time from the height; and this consists in finding that the first of these two quantities, according to the law which has been established, is a known multiple of the second power of the other; from which, finally, we have to calculate the value of the one when that of the other is given.

In this example the concrete question is more difficult than the abstract one. The reverse would be the case if we considered the same phenomenon in its greatest generality, as I have done above for another object. According to the circumstances, sometimes the first, sometimes the second, of these two parts will constitute the principal difficulty of the whole question; for the mathematical law of the phenomenon may be very simple, but very difficult to obtain, or it may be easy to discover, but very complicated; so that the two great sections of mathematical science, when we compare them as wholes, must be regarded as exactly equivalent in extent and in difficulty, as well as in importance, as we shall show farther on, in considering each of them separately.

Their different Natures. These two parts, essentially distinct in their *object*, as we have just seen, are no less so with regard to the *nature* of the inquiries of which they are composed.

The first should be called *concrete*, since it evidently depends on the character of the phenomena considered, and must necessarily vary when we examine new phenomena; while the second is completely independent of the nature of the objects examined, and is concerned with only the *numerical* relations which they present, for which reason it should be called *abstract*. The same relations may exist in a great number of different phenomena, which, in spite of their extreme diversity, will be viewed by the geometer as offering an analytical question susceptible, when studied by itself, of being resolved once for all. Thus, for instance, the same law which exists between the space and the time of the vertical fall of a body in a vacuum, is found again in many other phenomena which offer no analogy with the first nor with each other; for it expresses the relation between the surface of a spherical body and the length of its diameter; it determines, in like manner, the decrease of the intensity of light or of heat in relation to the distance of the objects lighted or heated, &c. The abstract part, common to these different mathematical questions, having been treated in reference to one of these, will thus have been treated for all; while the concrete part will have necessarily to be again taken up for each question separately, without the solution of any one of them being able to give any direct aid, in that connexion, for the solution of the rest.

The abstract part of mathematics is, then, general in its nature; the concrete part, special.

To present this comparison under a new point of view, we may say concrete mathematics has a philosophical character, which is essentially experimental, physical, phenomenal; while that of abstract mathematics is purely logical, rational. The concrete part of every mathematical question is necessarily founded on the consideration of the external world, and could never be resolved by a simple series of intellectual combinations. The abstract part, on the contrary, when it has been very completely separated, can consist only of a series of logical deductions, more or less prolonged; for if we have once found the equations of a phenomenon, the determination of the quantities therein considered, by means of one another, is a matter for reasoning only, whatever the difficulties may be. It belongs to the understanding alone to deduce from these equations results which are evidently contained in them, although perhaps in a very involved manner, without there being occasion to consult anew the external world; the consideration of which, having become thenceforth foreign to the subject, ought even to be carefully set aside in order to reduce the labour to its true peculiar difficulty. The *abstract* part of mathematics is then purely instrumental, and is only an immense and admirable extension of natural logic to a certain class of deductions. On the other hand, geometry and mechanics, which, as we shall see presently, constitute the *concrete* part, must be viewed as real natural sciences, founded on observation, like all the rest, although the extreme simplicity of their phenomena permits an infinitely greater degree of systematization, which has sometimes caused a misconception of the experimental character of their first principles.

We see, by this brief general comparison, how natural and profound is our fundamental division of mathematical science.

We have now to circumscribe, as exactly as we can in this first sketch, each of these two great sections.

CONCRETE MATHEMATICS.

Concrete Mathematics having for its object the discovery of the *equations* of phenomena, it would seem at first that it must be composed of as many distinct sciences as we find really distinct categories among natural phenomena. But we are yet very far from having discovered mathematical laws in all kinds of phenomena; we shall even see, presently, that the greater part will very probably always hide themselves from our investigations. In reality, in the present condition of the human mind, there are directly but two great general classes of phenomena, whose equations we constantly know; these are, firstly, geometrical, and, secondly, mechanical phenomena. Thus, then, the concrete part of mathematics is composed of GEOMETRY and RATIONAL MECHANICS.

This is sufficient, it is true, to give to it a complete character of logical universality, when we consider all phenomena from the most elevated point of view of natural philosophy. In fact, if all the parts of the universe were conceived as immovable, we should evidently have only geometrical phenomena to observe, since all would be reduced to relations of form, magnitude, and position; then, having regard to the motions which take place in it, we would have also to consider mechanical phenomena. Hence the universe, in the statical point of view, presents only geometrical phenomena; and, considered dynamically, only mechanical phenomena. Thus geometry and mechanics constitute the two fundamental natural sciences, in this sense, that all natural effects may be conceived as simple necessary results, either of the laws of extension or of the laws of motion.

But although this conception is always logically possible, the difficulty is to specialize it with the necessary precision, and to follow it exactly in each of the general cases offered to us by the study of nature; that is, to effectually reduce each principal question of natural philosophy, for a certain determinate order of phenomena, to the question of geometry or mechanics, to which we might rationally suppose it should be brought. This transformation, which requires great progress to have been previously made in the study of each class of phenomena, has thus far been really executed only for those of astronomy, and for a part of those considered by terrestrial physics, properly so called. It is thus that astronomy, acoustics, optics, &c., have finally become applications of mathematical science to certain orders of observations.* But these applications not being by their nature rigorously circumscribed, to confound them with the science would be to assign to it a vague and indefinite domain ; and this is done in the usual division, so faulty in so many other respects, of the mathematics into "Pure" and "Applied."

ABSTRACT MATHEMATICS.

The nature of abstract mathematics (the general division of which will be examined in the following chapter) is clearly and exactly determined. It is composed of what is called the *Calculus*,† taking this word in its greatest extent, which reaches from the most simple numerical operations to the most sublime combinations of transcendental analysis. The *Calculus* has the solution of all questions

---

* The investigation of the mathematical phenomena of the laws of heat by Baron Fourier has led to the establishment, in an entirely direct manner, of Thermological equations. This great discovery tends to elevate our philosophical hopes as to the future extensions of the legitimate applications of mathematical analysis, and renders it proper, in the opinion of the author, to regard *Thermology* as a third principal branch of concrete mathematics.

† The translator has felt justified in employing this very convenient word (for which our language has no precise equivalent) as an English one, in its most extended sense, in spite of its being often popularly confounded with its Differential and Integral department.

relating to numbers for its peculiar object. Its *starting point* is, constantly and necessarily, the knowledge of the precise relations, *i. e.*, of the *equations*, between the different magnitudes which are simultaneously considered; that which is, on the contrary, the *stopping point* of concrete mathematics. However complicated, or however indirect these relations may be, the final object of the calculus always is to obtain from them the values of the unknown quantities by means of those which are known. This *science*, although nearer perfection than any other, is really little advanced as yet, so that this object is rarely attained in a manner completely satisfactory.

Mathematical analysis is, then, the true rational basis of the entire system of our actual knowledge. It constitutes the first and the most perfect of all the fundamental sciences. The ideas with which it occupies itself are the most universal, the most abstract, and the most simple which it is possible for us to conceive.

This peculiar nature of mathematical analysis enables us easily to explain why, when it is properly employed, it is such a powerful instrument, not only to give more precision to our real knowledge, which is self-evident, but especially to establish an infinitely more perfect co-ordination in the study of the phenomena which admit of that application; for, our conceptions having been so generalized and simplified that a single analytical question, abstractly resolved, contains the *implicit* solution of a great number of diverse physical questions, the human mind must necessarily acquire by these means a greater facility in perceiving relations between phenomena which at first appeared entirely distinct from one another. We thus naturally see arise, through the medium of analysis, the most frequent and the most unexpected approximations between problems which at first offered no apparent connection, and which we often end in viewing as identical. Could we, for example, without the aid of analysis, perceive the least resemblance between the determination of the direction of a curve at each of its points and that of the velocity acquired by a body at every instant of its variable motion? and yet these questions, however different they may be, compose but one in the eyes of the geometer.

The high relative perfection of mathematical analysis is as easily perceptible. This perfection is not due, as some have thought, to the nature of the signs which are employed as instruments of reasoning, eminently concise and general as they are. In reality, all great analytical ideas have been formed without the algebraic signs having been of any essential aid, except for working them out after the mind had conceived them. The superior perfection of the science of the calculus is due principally to the extreme simplicity of the ideas which it considers, by whatever signs they may be expressed; so that there is not the least hope, by any artifice of scientific language, of perfecting to the same degree theories which refer to more complex subjects, and which are necessarily condemned by their nature to a greater or less logical inferiority.

THE EXTENT OF ITS FIELD.

Our examination of the philosophical character of mathematical science would remain incomplete, if, after having viewed its object and composition, we did not examine the real extent of its domain.

Its Universality. For this purpose it is indispensable to perceive, first of all, that, in the purely logical point of view, this science is by itself necessarily and rigorously universal; for there is no question whatever which may not be finally conceived as consisting in determining certain quantities from others by means of certain relations, and consequently as admitting of reduction, in final analysis, to a simple question of numbers. In all our researches, indeed, on whatever subject, our object is to arrive at numbers, at quantities, though often in a very imperfect manner and by very uncertain methods. Thus, taking an example in the class of subjects the least accessible to mathematics, the phenomena of living bodies, even when considered (to take the most complicated case) in the state of disease, is it not manifest that all the questions of therapeutics may be viewed as consisting in determining the *quantities* of the different agents which modify the organism, and which must act upon it to bring it to its normal state, admitting, for some of these quantities in certain cases, values which are equal to zero, or negative, or even contradictory?

The fundamental idea of Descartes on the relation of the concrete to the abstract in mathematics, has proven, in opposition to the superficial distinction of metaphysics, that all ideas of quality may be reduced to those of quantity. This conception, established at first by its immortal author in relation to geometrical phenomena only, has since been effectually extended to mechanical phenomena, and in our days to those of heat. As a result of this gradual generalization, there are now no geometers who do not consider it, in a purely theoretical sense, as capable of being applied to all our real ideas of every sort, so that every phenomenon is logically susceptible of being represented by an *equation;* as much so, indeed, as is a curve or a motion, excepting the difficulty of discovering it, and then of *resolving* it, which may be, and oftentimes are, superior to the greatest powers of the human mind.

Its Limitations. Important as it is to comprehend the rigorous universality, in a logical point of view, of mathematical science, it is no less indispensable to consider now the great real *limitations,* which, through the feebleness of our intellect, narrow in a remarkable degree its actual domain, in proportion as phenomena, in becoming special, become complicated.

Every question may be conceived as capable of being reduced to a pure question of numbers; but the difficulty of effecting such a transformation increases so much with the complication of the phenomena of natural philosophy, that it soon becomes insurmountable.

This will be easily seen, if we consider that to bring a question within the field of mathematical analysis, we must first have discovered the precise relations which exist between the quantities which are found in the phenomenon under examination, the establishment of these equations being the necessary starting point of all analytical labours. This must evidently be so much the more difficult as we have to do with phenomena which are more special, and therefore more complicated. We shall thus find that it is only in *inorganic physics,* at the most, that we can justly hope ever to obtain that high degree of scientific perfection.

The *first* condition which is necessary in order that phenomena may admit of mathematical laws, susceptible of being discovered, evidently is, that their different quantities should admit of being expressed by fixed numbers. We soon find that in this respect the whole of *organic physics*, and probably also the most complicated parts of inorganic physics, are necessarily inaccessible, by their nature, to our mathematical analysis, by reason of the extreme numerical variability of the corresponding phenomena. Every precise idea of fixed numbers is truly out of place in the phenomena of living bodies, when we wish to employ it otherwise than as a means of relieving the attention, and when we attach any importance to the exact relations of the values assigned.

We ought not, however, on this account, to cease to conceive all phenomena as being necessarily subject to mathematical laws, which we are condemned to be ignorant of, only because of the too great complication of the phenomena. The most complex phenomena of living bodies are doubtless essentially of no other special nature than the simplest phenomena of unorganized matter. If it were possible to isolate rigorously each of the simple causes which concur in producing a single physiological phenomenon, every thing leads us to believe that it would show itself endowed, in determinate circumstances, with a kind of influence and with a quantity of action as exactly fixed as we see it in universal gravitation, a veritable type of the fundamental laws of nature.

There is a *second* reason why we cannot bring complicated phenomena under the dominion of mathematical analysis. Even if we could ascertain the mathematical law which governs each agent, taken by itself, the combination of so great a number of conditions would render the corresponding mathematical problem so far above our feeble means, that the question would remain in most cases incapable of solution.

To appreciate this difficulty, let us consider how complicated mathematical questions become, even those relating to the most simple phenomena of unorganized bodies, when we desire to bring sufficiently near together the abstract and the concrete state, having regard to all the principal conditions which can exercise a real influence over the effect produced. We know, for example, that the very simple phenomenon of the flow of a fluid through a given orifice, by virtue of its gravity alone, has not as yet any complete mathematical solution, when we take into the account all the essential circumstances. It is the same even with the still more simple motion of a solid projectile in a resisting medium.

Why has mathematical analysis been able to adapt itself with such admirable success to the most profound study of celestial phenomena? Because they are, in spite of popular appearances, much more simple than any others. The most complicated problem which they present, that of the modification produced in the motions of two bodies tending towards each other by virtue of their gravitation, by the influence of a third body acting on both of them in the same manner, is much less complex than the most simple terrestrial problem. And, nevertheless, even it presents difficulties so great that we yet possess only approximate solutions of it. It is even easy to see that the high perfection to which solar astronomy has been able to elevate itself by the employment of mathematical science is, besides, essentially due to our having skilfully profited by all the particular, and, so to say, accidental facilities presented by the peculiarly favourable constitution of our planetary system. The planets which compose it are quite few in number, and their masses are in general very unequal, and much less than that of the sun; they are, besides, very distant from one another; they have forms almost spherical; their orbits are nearly circular, and only slightly inclined to each other, and so on. It results from all these circumstances that the perturbations are generally inconsiderable, and that to calculate them it is usually sufficient to take into the account, in connexion with the action of the sun on each particular planet, the influence of only one other planet, capable, by its size and its proximity, of causing perceptible derangements.

If, however, instead of such a state of things, our solar system had been composed of a greater number of planets concentrated into a less space, and nearly equal in mass; if their orbits had presented very different inclinations, and considerable eccentricities; if these bodies had been of a more complicated form, such as very eccentric ellipsoids, it is certain that, supposing the same law of gravitation to exist, we should not yet have succeeded in subjecting the study of the celestial phenomena to our mathematical analysis, and probably we should not even have been able to disentangle the present principal law.

These hypothetical conditions would find themselves exactly realized in the highest degree in *chemical* phenomena, if we attempted to calculate them by the theory of general gravitation.

On properly weighing the preceding considerations, the reader will be convinced, I think, that in reducing the future extension of the great applications of mathematical analysis, which are really possible, to the field comprised in the different departments of inorganic physics, I have rather exaggerated than contracted the extent of its actual domain. Important as it was to render apparent the rigorous logical universality of mathematical science, it was equally so to indicate the conditions which limit for us its real extension, so as not to contribute to lead the human mind astray from the true scientific direction in the study of the most complicated phenomena, by the chimerical search after an impossible perfection.

Having thus exhibited the essential object and the principal composition of mathematical science, as well as its general relations with the whole body of natural philosophy, we have now to pass to the special examination of the great sciences of which it is composed.

Note.—ANALYSIS and GEOMETRY are the two great heads under which the subject is about to be examined. To these *M. Comte* adds Rational MECHANICS; but as it is not comprised in the usual idea of Mathematics, and as its discussion would be of but limited utility and interest, it is not included in the present translation.

BOOK I.

ANALYSIS.

BOOK I.

ANALYSIS.

CHAPTER I.

GENERAL VIEW OF MATHEMATICAL ANALYSIS.

In the historical development of mathematical science since the time of Descartes, the advances of its abstract portion have always been determined by those of its concrete portion; but it is none the less necessary, in order to conceive the science in a manner truly logical, to consider the Calculus in all its principal branches before proceeding to the philosophical study of Geometry and Mechanics. Its analytical theories, more simple and more general than those of concrete mathematics, are in themselves essentially independent of the latter; while these, on the contrary, have, by their nature, a continual need of the former, without the aid of which they could make scarcely any progress. Although the principal conceptions of analysis retain at present some very perceptible traces of their geometrical or mechanical origin, they are now, however, mainly freed from that primitive character, which no longer manifests itself except in some secondary points; so that it is possible (especially since the labours of Lagrange) to present them in a dogmatic exposition, by a purely abstract method, in a single and continuous system. It is this which will be undertaken in the present and the five following chapters, limiting our investigations to the most general considerations upon each principal branch of the science of the calculus.

The definite object of our researches in concrete mathematics being the discovery of the *equations* which express the mathematical laws of the phenomenon under consideration, and these equations constituting the true starting point of the calculus, which has for its object to obtain from them the determination of certain quantities by means of others, I think it indispensable, before proceeding any farther, to go more deeply than has been customary into that fundamental idea of *equation*, the continual subject, either as end or as beginning, of all mathematical labours. Besides the advantage of circumscribing more definitely the true field of analysis, there will result from it the important consequence of tracing in a more exact manner the real line of demarcation between the concrete and the abstract part of mathematics, which will complete the general exposition of the fundamental division established in the introductory chapter.

THE TRUE IDEA OF AN EQUATION.

We usually form much too vague an idea of what an *equation* is, when we give that name to every kind of relation of equality between *any* two functions of the magnitudes which we are considering. For, though every equation is evidently a relation of equality, it is far from being true that, reciprocally, every relation of equality is a veritable *equation*, of the kind of those to which, by their nature, the methods of analysis are applicable.

This want of precision in the logical consideration of an idea which is so fundamental in mathematics, brings with it the serious inconvenience of rendering it almost impossible to explain, in general terms, the great and fundamental difficulty which we find in establishing the relation between the concrete and the abstract, and which stands out so prominently in each great mathematical question taken by itself. If the meaning of the word *equation* was truly as extended as we habitually suppose it to be in our definition of it, it is not apparent what great difficulty there could really be, in general, in establishing the equations of any problem whatsoever; for the whole would thus appear to consist in a simple question of form, which ought never even to exact any great intellectual efforts, seeing that we can hardly conceive of any precise relation which is not immediately a certain relation of equality, or which cannot be readily brought thereto by some very easy transformations.

Thus, when we admit every species of *functions* into the definition of *equations*, we do not at all account for the extreme difficulty which we almost always experience in putting a problem into an equation, and which so often may be compared to the efforts required by the analytical elaboration of the equation when once obtained. In a word, the ordinary abstract and general idea of an *equation* does not at all correspond to the real meaning which geometers attach to that expression in the actual development of the science. Here, then, is a logical fault, a defect of correlation, which it is very important to rectify.

Division of Functions into Abstract and Concrete. To succeed in doing so, I begin by distinguishing two sorts of *functions*, *abstract* or analytical functions, and *concrete* functions. The first alone can enter into veritable *equations*. We may, therefore, henceforth define every *equation*, in an exact and sufficiently profound manner, as a relation of equality between two *abstract* functions of the magnitudes under consideration. In order not to have to return again to this fundamental definition, I must add here, as an indispensable complement, without which the idea would not be sufficiently general, that these abstract functions may refer not only to the magnitudes which the problem presents of itself, but also to all the other auxiliary magnitudes which are connected with it, and which we will often be able to introduce, simply as a mathematical artifice, with the sole object of facilitating the discovery of the equations of the phenomena. I here anticipate summarily the result of a general discussion of the highest importance, which will be found at the end of this chapter. We will now return to the essential distinction of functions as abstract and concrete.

This distinction may be established in two ways, essentially different, but complementary of each other, *à priori* and *à posteriori*; that is to say, by characterizing in a general manner the peculiar nature of each species of functions, and then by making the actual enumeration of all the abstract functions at present known, at least so far as relates to the elements of which they are composed.

A priori, the functions which I call *abstract* are those which express a manner of dependence between magnitudes, which can be conceived between numbers alone, without there being need of indicating any phenomenon whatever in which it is realized. I name, on the other hand, *concrete* functions, those for which the mode of dependence expressed cannot be defined or conceived except by assigning a determinate case of physics, geometry, mechanics, &c., in which it actually exists.

Most functions in their origin, even those which are at present the most purely *abstract*, have begun by being *concrete;* so that it is easy to make the preceding distinction understood, by citing only the successive different points of view under which, in proportion as the science has become formed, geometers have considered the most simple analytical functions. I will indicate powers, for example, which have in general become abstract functions only since the labours of Vieta and Descartes. The functions x^2, x^3, which in our present analysis are so well conceived as simply *abstract*, were, for the geometers of antiquity, perfectly *concrete* functions, expressing the relation of the superficies of a square, or the volume of a cube to the length of their side. These had in their eyes such a character so exclusively, that it was only by means of the geometrical definitions that they discovered the elementary algebraic properties of these functions, relating to the decomposition of the variable into two parts, properties which were at that epoch only real theorems of geometry, to which a numerical meaning was not attached until long afterward.

I shall have occasion to cite presently, for another reason, a new example, very suitable to make apparent the fundamental distinction which I have just exhibited; it is that of circular functions, both direct and inverse, which at the present time are still sometimes concrete, sometimes abstract, according to the point of view under which they are regarded.

A posteriori, the general character which renders a function abstract or concrete having been established, the question as to whether a certain determinate function is veritably abstract, and therefore susceptible of entering into true analytical equations, becomes a simple question of fact, inasmuch as we are going to enumerate all the functions of this species.

Enumeration of Abstract Functions. At first view this enumeration seems impossible, the distinct analytical functions being infinite in number. But when we divide them into *simple* and *compound*, the difficulty disappears; for, though the number of the different functions considered in mathematical analysis is really infinite, they are, on the contrary, even at the present day, composed of a very small number of elementary functions, which can be easily assigned, and which are evidently sufficient for deciding the abstract or concrete character of any given function; which will be of the one or the other nature, according as it shall be composed exclusively of these simple abstract functions, or as it shall include others.

We evidently have to consider, for this purpose, only the functions of a single variable, since those relative to several independent variables are constantly, by their nature, more or less *compound.*

Let x be the independent variable, y the correlative variable which depends upon it. The different simple modes of abstract dependence, which we can now conceive between y and x, are expressed by the ten following elementary formulas, in which each function is coupled with its *inverse*, that is, with that which would be obtained from the direct function by referring x to y, instead of referring y to x.

		FUNCTION.	ITS NAME.
1st couple	1°	$y=a+x$	*Sum.*
	2°	$y=a-x$	*Difference.*
2d couple	1°	$y=ax$	*Product.*
	2°	$y=\frac{a}{x}$	*Quotient.*
3d couple	1°	$y=x^a$	*Power.*
	2°	$y=\sqrt[a]{x}$	*Root.*
4th couple	1°	$y=a^x$	*Exponential.*
	2°	$y=l_a x$	*Logarithmic.*
5th couple	1°	$y=\sin. x$	*Direct Circular.*
	2°	$y=\text{arc}(\sin.=x)$.	*Inverse Circular.**

Such are the elements, very few in number, which directly compose all the abstract functions known at the present day. Few as they are, they are evidently sufficient to give rise to an infinite number of analytical combinations.

---

* With the view of increasing as much as possible the resources and the extent (now so insufficient) of mathematical analysis, geometers count this last couple of functions among the analytical elements. Although this inscription is strictly legitimate, it is important to remark that circular functions are not exactly in the same situation as the other abstract elementary functions. There is this very essential difference, that the functions of the four first couples are at the same time simple and abstract, while the circular functions, which may manifest each character in succession, according to the point of view under which they are considered and the manner in which they are employed, never present these two properties simultaneously.

Some other concrete functions may be usefully introduced into the number of analytical elements, certain conditions being fulfilled. It is thus, for example, that the labours of M. Legendre and of M. Jacobi on *elliptical* functions have truly enlarged the field of analysis; and the same is true of some definite integrals obtained by M. Fourier in the theory of heat.

No rational consideration rigorously circumscribes, *à priori*, the preceding table, which is only the actual expression of the present state of the science. Our analytical elements are at the present day more numerous than they were for Descartes, and even for Newton and Leibnitz: it is only a century since the last two couples have been introduced into analysis by the labours of John Bernouilli and Euler. Doubtless new ones will be hereafter admitted; but, as I shall show towards the end of this chapter, we cannot hope that they will ever be greatly multiplied, their real augmentation giving rise to very great difficulties.

We can now form a definite, and, at the same time, sufficiently extended idea of what geometers understand by a veritable *equation*. This explanation is especially suited to make us understand how difficult it must be really to establish the *equations* of phenomena, since we have effectually succeeded in so doing only when we have been able to conceive the mathematical laws of these phenomena by the aid of functions entirely composed of only the mathematical elements which I have just enumerated. It is clear, in fact, that it is then only that the problem becomes truly abstract, and is reduced to a pure question of numbers, these functions being the only simple relations which we can conceive between numbers, considered by themselves. Up to this period of the solution, whatever the appearances may be, the question is still essentially concrete, and does not come within the domain of the *calculus*. Now the fundamental difficulty of this passage from the *concrete* to the *abstract* in general consists especially in the insufficiency of this very small number of analytical elements which we possess, and by means of which, nevertheless, in spite of the little real variety which they offer us, we must succeed in representing all the precise relations which all the different natural phenomena can manifest to us. Considering the infinite diversity which must necessarily exist in this respect in the external world, we easily understand how far below the true difficulty our conceptions must frequently be found, especially if we add that as these elements of our analysis have been in the first place furnished to us by the mathematical consideration of the simplest phenomena, we have, *à priori*, no rational guarantee of their necessary suitableness to represent the mathematical law of every other class of phenomena. I will explain presently the general artifice, so profoundly ingenious, by which the human mind has succeeded in diminishing, in a remarkable degree, this fundamental difficulty which is presented by the relation of the concrete to the abstract in mathematics, without, however, its having been necessary to multiply the number of these analytical elements.

THE TWO PRINCIPAL DIVISIONS OF THE CALCULUS.

The preceding explanations determine with precision the true object and the real field of abstract mathematics. I must now pass to the examination of its principal divisions, for thus far we have considered the calculus as a whole.

The first direct consideration to be presented on the composition of the science of the *calculus* consists in dividing it, in the first place, into two principal branches, to which, for want of more suitable denominations, I will give the names of *Algebraic calculus*, or *Algebra*, and of

Arithmetical calculus, or *Arithmetic;* but with the caution to take these two expressions in their most extended logical acceptation, in the place of the by far too restricted meaning which is usually attached to them.

The complete solution of every question of the *calculus*, from the most elementary up to the most transcendental, is necessarily composed of two successive parts, whose nature is essentially distinct. In the first, the object is to transform the proposed equations, so as to make apparent the manner in which the unknown quantities are formed by the known ones: it is this which constitutes the *algebraic* question. In the second, our object is to *find the values* of the formulas thus obtained; that is, to determine directly the values of the numbers sought, which are already represented by certain explicit functions of given numbers: this is the *arithmetical* question.* It is apparent that, in every solution which is

* Suppose, for example, that a question gives the following equation between an unknown magnitude x, and two known magnitudes, a and b,

$$x^3+3ax=2b,$$

as is the case in the problem of the trisection of an angle. We see at once that the dependence between x on the one side, and ab on the other, is completely determined; but, so long as the equation preserves its primitive form, we do not at all perceive in what manner the unknown quantity is derived from the data. This must be discovered, however, before we can think of determining its value. Such is the object of the algebraic part of the solution. When, by a series of transformations which have successively rendered that derivation more and more apparent, we have arrived at presenting the proposed equation under the form

$$x=\sqrt[3]{b+\sqrt{b^2+a^3}}+\sqrt[3]{b-\sqrt{b^2+a^3}},$$

the work of *algebra* is finished; and even if we could not perform the arithmetical operations indicated by that formula, we would nevertheless have obtained a knowledge very real, and often very important. The work of *arithmetic* will now consist in taking that formula for its starting point, and finding the number x when the values of the numbers a and b are given.

truly rational, it necessarily follows the algebraical question, of which it forms the indispensable complement, since it is evidently necessary to know the mode of generation of the numbers sought for before determining their actual values for each particular case. Thus the stopping-place of the algebraic part of the solution becomes the starting point of the arithmetical part.

We thus see that the *algebraic* calculus and the *arithmetical* calculus differ essentially in their object. They differ no less in the point of view under which they regard quantities; which are considered in the first as to their *relations*, and in the second as to their *values*. The true spirit of the calculus, in general, requires this distinction to be maintained with the most severe exactitude, and the line of demarcation between the two periods of the solution to be rendered as clear and distinct as the proposed question permits. The attentive observation of this precept, which is too much neglected, may be of much assistance, in each particular question, in directing the efforts of our mind, at any moment of the solution, towards the real corresponding difficulty. In truth, the imperfection of the science of the calculus obliges us very often (as will be explained in the next chapter) to intermingle algebraic and arithmetical considerations in the solution of the same question. But, however impossible it may be to separate clearly the two parts of the labour, yet the preceding indications will always enable us to avoid confounding them.

In endeavouring to sum up as succinctly as possible the distinction just established, we see that ALGEBRA may be defined, in general, as having for its object the *resolution of equations;* taking this expression in its full logical meaning, which signifies the transformation of *implicit* functions into equivalent *explicit* ones. In the same way, ARITHMETIC may be defined as destined to *the determination of the values of functions*. Henceforth, therefore, we will briefly say that ALGEBRA is the *Calculus of Functions*, and ARITHMETIC the *Calculus of Values*.

We can now perceive how insufficient and even erroneous are the ordinary definitions. Most generally, the exaggerated importance attributed to Signs has led to the distinguishing the two fundamental branches of the science of the Calculus by the manner of designating in each the subjects of discussion, an idea which is evidently absurd in principle and false in fact. Even the celebrated definition given by Newton, characterizing *Algebra* as *Universal Arithmetic*, gives certainly a very false idea of the nature of algebra and of that of arithmetic.*

Having thus established the fundamental division of the calculus into two principal branches, I have now to compare in general terms the extent, the importance, and the difficulty of these two sorts of calculus, so as to have hereafter to consider only the *Calculus of Functions*, which is to be the principal subject of our study.

---

* I have thought that I ought to specially notice this definition, because it serves as the basis of the opinion which many intelligent persons, unacquainted with mathematical science, form of its abstract part, without considering that at the time of this definition mathematical analysis was not sufficiently developed to enable the general character of each of its principal parts to be properly apprehended, which explains why Newton could at that time propose a definition which at the present day he would certainly reject.

THE CALCULUS OF VALUES, OR ARITHMETIC.

Its Extent. The *Calculus of Values, or Arithmetic,* would appear, at first view, to present a field as vast as that of *algebra,* since it would seem to admit as many distinct questions as we can conceive different algebraic formulas whose values are to be determined. But a very simple reflection will show the difference. Dividing functions into *simple* and *compound,* it is evident that when we know how to determine the *value* of simple functions, the consideration of compound functions will no longer present any difficulty. In the algebraic point of view, a compound function plays a very different part from that of the elementary functions of which it consists, and from this, indeed, proceed all the principal difficulties of analysis. But it is very different with the Arithmetical Calculus. Thus the number of truly distinct arithmetical operations is only that determined by the number of the elementary abstract functions, the very limited list of which has been given above. The determination of the values of these ten functions necessarily gives that of all the functions, infinite in number, which are considered in the whole of mathematical analysis, such at least as it exists at present. There can be no new arithmetical operations without the creation of really new analytical elements, the number of which must always be extremely small. The field of *arithmetic* is, then, by its nature, exceedingly restricted, while that of algebra is rigorously indefinite.

It is, however, important to remark, that the domain of the *calculus of values* is, in reality, much more extensive than it is commonly represented; for several questions truly *arithmetical*, since they consist of determinations of values, are not ordinarily classed as such, because we are accustomed to treat them only as incidental in the midst of a body of analytical researches more or less elevated, the too high opinion commonly formed of the influence of signs being again the principal cause of this confusion of ideas. Thus not only the construction of a table of logarithms, but also the calculation of trigonometrical tables, are true arithmetical operations of a higher kind. We may also cite as being in the same class, although in a very distinct and more elevated order, all the methods by which we determine directly the value of any function for each particular system of values attributed to the quantities on which it depends, when we cannot express in general terms the explicit form of that function. In this point of view the *numerical* solution of questions which we cannot resolve algebraically, and even the calculation of "Definite Integrals," whose general integrals we do not know, really make a part, in spite of all appearances, of the domain of *arithmetic*, in which we must necessarily comprise all that which has for its object the *determination of the values of functions*. The considerations relative to this object are, in fact, constantly homogeneous, whatever the *determinations* in question, and are always very distinct from truly *algebraic* considerations.

To complete a just idea of the real extent of the calculus of values, we must include in it likewise that part of the general science of the calculus which now bears the name of the *Theory of Numbers*, and which is yet so little advanced. This branch, very extensive by its nature, but whose importance in the general system of science is not very great, has for its object the discovery of the properties inherent in different numbers by virtue of their values, and independent of any particular system of numeration. It forms, then, a sort of *transcendental arithmetic;* and to it would really apply the definition proposed by Newton for algebra.

The entire domain of arithmetic is, then, much more extended than is commonly supposed; but this *calculus of values* will still never be more than a point, so to speak, in comparison with the *calculus of functions*, of which mathematical science essentially consists. This comparative estimate will be still more apparent from some considerations which I have now to indicate respecting the true nature of arithmetical questions in general, when they are more profoundly examined.

Its true Nature. In seeking to determine with precision in what *determinations of values* properly consist, we easily recognize that they are nothing else but veritable *transformations* of the functions to be valued; transformations which, in spite of their special end, are none the less essentially of the same nature as all those taught by analysis. In this point of view, the *calculus of values* might be simply conceived as an appendix, and a particular application of the *calculus of functions*, so that *arithmetic* would disappear, so to say, as a distinct section in the whole body of abstract mathematics.

In order thoroughly to comprehend this consideration, we must observe that, when we propose to determine the *value* of an unknown number whose mode of formation is given, it is, by the mere enunciation of the arithmetical question, already defined and expressed under a certain form; and that in *determining its value* we only put its expression under another determinate form, to which we are accustomed to refer the exact notion of each particular number by making it re-enter into the regular system of *numeration*. The determination of values consists so completely of a simple *transformation*, that when the primitive expression of the number is found to be already conformed to the regular system of numeration, there is no longer any determination of value, properly speaking, or, rather, the question is answered by the question itself. Let the question be to add the two numbers *one* and *twenty*, we answer it by merely repeating the enunciation of the question,* and nevertheless we think that we have *determined the value* of the sum. This signifies that in this case the first expression of the function had no need of being transformed, while it would not be thus in adding twenty-three and fourteen, for then the sum would not be immediately expressed in a manner conformed to the rank which it occupies in the fixed and general scale of numeration.

To sum up as comprehensively as possible the preceding views, we may say, that to determine the *value* of a number is nothing else than putting its primitive expression under the form

$$a+bz+cz^2+dz^3+ez^4 \; . \; . \; . \; . \; . \; +pz^m,$$

z being generally equal to 10, and the coefficients a, b, c, d, &c., being subjected to the conditions of being whole numbers less than z; capable of becoming equal to zero; but never negative. Every arithmetical question may thus be stated as consisting in putting under such a form

* This is less strictly true in the English system of numeration than in the French, since "twenty-one" is our more usual mode of expressing this number.

any abstract function whatever of different quantities, which are supposed to have themselves a similar form already. We might then see in the different operations of arithmetic only simple particular cases of certain algebraic transformations, excepting the special difficulties belonging to conditions relating to the nature of the coefficients.

It clearly follows that abstract mathematics is essentially composed of the *Calculus of Functions*, which had been already seen to be its most important, most extended, and most difficult part. It will henceforth be the exclusive subject of our analytical investigations. I will therefore no longer delay on the *Calculus of Values*, but pass immediately to the examination of the fundamental division of the *Calculus of Functions.*

THE CALCULUS OF FUNCTIONS, OR ALGEBRA.

Principle of its Fundamental Division. We have determined, at the beginning of this chapter, wherein properly consists the difficulty which we experience in putting mathematical questions into *equations*. It is essentially because of the insufficiency of the very small number of analytical elements which we possess, that the relation of the concrete to the abstract is usually so difficult to establish. Let us endeavour now to appreciate in a philosophical manner the general process by which the human mind has succeeded, in so great a number of important cases, in overcoming this fundamental obstacle to *The establishment of Equations.*

1. *By the Creation of new Functions.* In looking at this important question from the most general point of view, we are led at once to the conception of one means of

facilitating the establishment of the equations of phenomena. Since the principal obstacle in this matter comes from the too small number of our analytical elements, the whole question would seem to be reduced to creating new ones. But this means, though natural, is really illusory; and though it might be useful, it is certainly insufficient.

In fact, the creation of an elementary abstract function, which shall be veritably new, presents in itself the greatest difficulties. There is even something contradictory in such an idea; for a new analytical element would evidently not fulfil its essential and appropriate conditions, if we could not immediately *determine its value*. Now, on the other hand, how are we to *determine the value* of a new function which is truly *simple*, that is, which is not formed by a combination of those already known? That appears almost impossible. The introduction into analysis of another elementary abstract function, or rather of another couple of functions (for each would be always accompanied by its *inverse*), supposes then, of necessity, the simultaneous creation of a new arithmetical operation, which is certainly very difficult.

If we endeavour to obtain an idea of the means which the human mind employs for inventing new analytical elements, by the examination of the procedures by the aid of which it has actually conceived those which we already possess, our observations leave us in that respect in an entire uncertainty, for the artifices which it has already made use of for that purpose are evidently exhausted. To convince ourselves of it, let us consider the last couple of simple functions which has been introduced into analysis, and at the formation of which we have been present, so to speak, namely, the fourth couple; for, as I have explained, the fifth couple does not strictly give veritable new analytical elements. The function a^x, and, consequently, its inverse, have been formed by conceiving, under a new point of view, a function which had been a long time known, namely, powers—when the idea of them had become sufficiently generalized. The consideration of a power relatively to the variation of its exponent, instead of to the variation of its base, was sufficient to give rise to a truly novel simple function, the variation following then an entirely different route. But this artifice, as simple as ingenious, can furnish nothing more; for, in turning over in the same manner all our present analytical elements, we end in only making them return into one another.

We have, then, no idea as to how we could proceed to the creation of new elementary abstract functions which would properly satisfy all the necessary conditions. This is not to say, however, that we have at present attained the effectual limit established in that respect by the bounds of our intelligence. It is even certain that the last special improvements in mathematical analysis have contributed to extend our resources in that respect, by introducing within the domain of the calculus certain definite integrals, which in some respects supply the place of new simple functions, although they are far from fulfilling all the necessary conditions, which has prevented me from inserting them in the table of true analytical elements. But, on the whole, I think it unquestionable that the number of these elements cannot increase except with extreme slowness. It is therefore not from these sources that the human mind has drawn its most powerful means of facilitating, as much as is possible, the establishment of equations.

2. *By the Conception of Equations between certain auxiliary Quantities.* This first method being set aside, there remains evidently but one other: it is, seeing the impossibility of finding directly the equations between the quantities under consideration, to seek for corresponding ones between other auxiliary quantities, connected with the first according to a certain determinate law, and from the relation between which we may return to that between the primitive magnitudes. Such is, in substance, the eminently fruitful conception which the human mind has succeeded in establishing, and which constitutes its most admirable instrument for the mathematical explanation of natural phenomena; the *analysis*, called *transcendental.*

As a general philosophical principle, the auxiliary quantities, which are introduced in the place of the primitive magnitudes, or concurrently with them, in order to facilitate the establishment of equations, might be derived according to any law whatever from the immediate elements of the question. This conception has thus a much more extensive reach than has been commonly attributed to it by even the most profound geometers. It is extremely important for us to view it in its whole logical extent, for it will perhaps be by establishing a general mode of *derivation* different from that to which we have thus far confined ourselves (although it is evidently very far from being the only possible one) that we shall one day succeed in essentially perfecting mathematical analysis as a whole, and consequently in establishing more powerful means of investigating the laws of nature than our present processes, which are unquestionably susceptible of becoming exhausted.

But, regarding merely the present constitution of the science, the only auxiliary quantities habitually introduced in the place of the primitive quantities in the *Transcendental Analysis* are what are called, 1°, *infinitely small* elements, the *differentials* (of different orders) of those quantities, if we regard this analysis in the manner of LEIBNITZ; or, 2°, the *fluxions*, the limits of the ratios of the simultaneous increments of the primitive quantities compared with one another, or, more briefly, the *prime and ultimate ratios* of these increments, if we adopt the conception of NEWTON; or, 3°, the *derivatives*, properly so called, of those quantities, that is, the coefficients of the different terms of their respective increments, according to the conception of LAGRANGE.

These three principal methods of viewing our present transcendental analysis, and all the other less distinctly characterized ones which have been successively proposed, are, by their nature, necessarily identical, whether in the calculation or in the application, as will be explained in a general manner in the third chapter. As to their relative value, we shall there see that the conception of Leibnitz has thus far, in practice, an incontestable superiority, but that its logical character is exceedingly vicious; while that the conception of Lagrange, admirable by its simplicity, by its logical perfection, by the philosophical unity which it has established in mathematical analysis (till then separated into two almost entirely independent worlds), presents, as yet, serious inconveniences in the applications, by retarding the progress of the mind. The conception of Newton occupies nearly middle ground in these various relations, being less rapid, but more rational than that of Leibnitz; less philosophical, but more applicable than that of Lagrange.

This is not the place to explain the advantages of the introduction of this kind of auxiliary quantities in the place of the primitive magnitudes. The third chapter is devoted to this subject. At present I limit myself to consider this conception in the most general manner, in order to deduce therefrom the fundamental division of the *calculus of functions* into two systems essentially distinct, whose dependence, for the complete solution of any one mathematical question, is invariably determinate.

In this connexion, and in the logical order of ideas, the transcendental analysis presents itself as being necessarily the first, since its general object is to facilitate the establishment of equations, an operation which must evidently precede the *resolution* of those equations, which is the object of the ordinary analysis. But though it is exceedingly important to conceive in this way the true relations of these two systems of analysis, it is none the less proper, in conformity with the regular usage, to study the transcendental analysis after ordinary analysis; for though the former is, at bottom, by itself logically independent of the latter, or, at least, may be essentially disengaged from it, yet it is clear that, since its employment in the solution of questions has always more or less need of being completed by the use of the ordinary analysis, we would be constrained to leave the questions in suspense if this latter had not been previously studied.

Corresponding Divisions of the Calculus of Functions. It follows from the preceding considerations that the *Calculus of Functions*, or *Algebra* (taking this word in its most extended meaning), is composed of two distinct fundamental branches, one of which has for its immediate object the *resolution* of equations, when they are directly established between the magnitudes themselves which are under consideration; and the other, starting from equations (generally much easier to form) between quantities indirectly connected with those of the problem, has for its peculiar and constant destination the deduction, by invariable analytical methods, of the corresponding equations between the direct magnitudes which we are considering; which brings the question within the domain of the preceding calculus.

The former calculus bears most frequently the name of *Ordinary Analysis*, or of *Algebra*, properly so called. The second constitutes what is called the *Transcendental Analysis*, which has been designated by the different denominations of *Infinitesimal Calculus*, *Calculus of Fluxions and of Fluents*, *Calculus of Vanishing Quantities*, the *Differential and Integral Calculus*, &c., according to the point of view in which it has been conceived.

In order to remove every foreign consideration, I will propose to name it CALCULUS OF INDIRECT FUNCTIONS, giving to ordinary analysis the title of CALCULUS OF DIRECT FUNCTIONS. These expressions, which I form essentially by generalizing and epitomizing the ideas of Lagrange, are simply intended to indicate with precision the true general character belonging to each of these two forms of analysis.

Having now established the fundamental division of mathematical analysis, I have next to consider separately each of its two parts, commencing with the *Calculus of Direct Functions*, and reserving more extended developments for the different branches of the *Calculus of Indirect Functions*.

CHAPTER II.

ORDINARY ANALYSIS, OR ALGEBRA.

THE *Calculus of direct Functions*, or *Algebra*, is (as was shown at the end of the preceding chapter) entirely sufficient for the solution of mathematical questions, when they are so simple that we can form directly the equations between the magnitudes themselves which we are considering, without its being necessary to introduce in their place, or conjointly with them, any system of auxiliary quantities *derived* from the first. It is true that in the greatest number of important cases its use requires to be preceded and prepared by that of the *Calculus of indirect Functions*, which is intended to facilitate the establishment of equations. But, although algebra has then only a secondary office to perform, it has none the less a necessary part in the complete solution of the question, so that the *Calculus of direct Functions* must continue to be, by its nature, the fundamental base of all mathematical analysis. We must therefore, before going any further, consider in a general manner the logical composition of this calculus, and the degree of development to which it has at the present day arrived.

Its Object. The final object of this calculus being the *resolution* (properly so called) of *equations*, that is, the discovery of the manner in which the unknown quantities are formed from the known quantities, in accordance with the *equations* which exist between them, it naturally presents as many different departments as we can conceive truly distinct classes of equations. Its appropriate extent is consequently rigorously indefinite, the number of analytical functions susceptible of entering into equations being in itself quite unlimited, although they are composed of only a very small number of primitive elements.

Classification of Equations. The rational classification of equations must evidently be determined by the nature of the analytical elements of which their numbers are composed; every other classification would be essentially arbitrary. Accordingly, analysts begin by dividing equations with one or more variables into two principal classes, according as they contain functions of only the first three couples (see the table in chapter i., page 51), or as they include also exponential or circular functions. The names of *Algebraic* functions and *Transcendental* functions, commonly given to these two principal groups of analytical elements, are undoubtedly very inappropriate. But the universally established division between the corresponding equations is none the less very real in this sense, that the resolution of equations containing the functions called *transcendental* necessarily presents more difficulties than those of the equations called *algebraic.* Hence the study of the former is as yet exceedingly imperfect, so that frequently the resolution of the most simple of them is still unknown to us,* and our analytical methods have almost exclusive reference to the elaboration of the latter.

---

* Simple as may seem, for example, the equation

$$a^x + b^x = c^x,$$

we do not yet know how to resolve it, which may give some idea of the extreme imperfection of this part of algebra.

ALGEBRAIC EQUATIONS.

Considering now only these *Algebraic* equations, we must observe, in the first place, that although they may often contain *irrational* functions of the unknown quantities as well as *rational* functions, we can always, by more or less easy transformations, make the first case come under the second, so that it is with this last that analysts have had to occupy themselves exclusively in order to resolve all sorts of *algebraic* equations.

Their Classification. In the infancy of algebra, these equations were classed according to the number of their terms. But this classification was evidently faulty, since it separated cases which were really similar, and brought together others which had nothing in common besides this unimportant characteristic.* It has been retained only for equations with two terms, which are, in fact, capable of being resolved in a manner peculiar to themselves.

The classification of equations by what is called their *degrees*, is, on the other hand, eminently natural, for this distinction rigorously determines the greater or less difficulty of their *resolution.* This gradation is apparent in the cases of all the equations which can be resolved; but it may be indicated in a general manner independently of the fact of the resolution. We need only consider that the most general equation of each degree necessarily comprehends all those of the different inferior degrees, as must also the formula which determines the unknown quantity. Consequently, however slight we may suppose the difficulty peculiar to the *degree* which we

* The same error was afterward committed, in the infancy of the infinitesimal calculus, in relation to the integration of differential equations.

are considering, since it is inevitably complicated in the execution with those presented by all the preceding degrees, the resolution really offers more and more obstacles, in proportion as the degree of the equation is elevated.

ALGEBRAIC RESOLUTION OF EQUATIONS.

Its Limits. The resolution of algebraic equations is as yet known to us only in the four first degrees, such is the increase of difficulty noticed above. In this respect, algebra has made no considerable progress since the labours of Descartes and the Italian analysts of the sixteenth century, although in the last two centuries there has been perhaps scarcely a single geometer who has not busied himself in trying to advance the resolution of equations. The general equation of the fifth degree itself has thus far resisted all attacks.

The constantly increasing complication which the formulas for resolving equations must necessarily present, in proportion as the degree increases (the difficulty of using the formula of the fourth degree rendering it almost inapplicable), has determined analysts to renounce, by a tacit agreement, the pursuit of such researches, although they are far from regarding it as impossible to obtain the resolution of equations of the fifth degree, and of several other higher ones.

General Solution. The only question of this kind which would be really of great importance, at least in its logical relations, would be the general resolution of algebraic equations of any degree whatsoever. Now, the more we meditate on this subject, the more we are led to think, with Lagrange, that it really surpasses the scope of our intelligence. We must besides observe that the formula which would express the *root* of an equation of the mth degree would necessarily include radicals of the mth order (or functions of an equivalent multiplicity), because of the m determinations which it must admit. Since we have seen, besides, that this formula must also embrace, as a particular case, that formula which corresponds to every lower degree, it follows that it would inevitably also contain radicals of the next lower degree, the next lower to that, &c., so that, even if it were possible to discover it, it would almost always present too great a complication to be capable of being usefully employed, unless we could succeed in simplifying it, at the same time retaining all its generality, by the introduction of a new class of analytical elements of which we yet have no idea. We have, then, reason to believe that, without having already here arrived at the limits imposed by the feeble extent of our intelligence, we should not be long in reaching them if we actively and earnestly prolonged this series of investigations.

It is, besides, important to observe that, even supposing we had obtained the resolution of *algebraic* equations of any degree whatever, we would still have treated only a very small part of *algebra*, properly so called, that is, of the calculus of direct functions, including the resolution of all the equations which can be formed by the known analytical functions.

Finally, we must remember that, by an undeniable law of human nature, our means for conceiving new questions being much more powerful than our resources for resolving them, or, in other words, the human mind being much more ready to inquire than to reason, we shall necessarily always remain *below* the difficulty, no matter to what degree of development our intellectual labour may arrive. Thus, even though we should some day discover the complete resolution of all the analytical equations at present known, chimerical as the supposition is, there can be no doubt that, before attaining this end, and probably even as a subsidiary means, we would have already overcome the difficulty (a much smaller one, though still very great) of conceiving new analytical elements, the introduction of which would give rise to classes of equations of which, at present, we are completely ignorant; so that a similar imperfection in algebraic science would be continually reproduced, in spite of the real and very important increase of the absolute mass of our knowledge.

What we know in Algebra. In the present condition of algebra, the complete resolution of the equations of the first four degrees, of any binomial equations, of certain particular equations of the higher degrees, and of a very small number of exponential, logarithmic, or circular equations, constitute the fundamental methods which are presented by the calculus of direct functions for the solution of mathematical problems. But, limited as these elements are, geometers have nevertheless succeeded in treating, in a truly admirable manner, a very great number of important questions, as we shall find in the course of the volume. The general improvements introduced within a century into the total system of mathematical analysis, have had for their principal object to make immeasurably useful this little knowledge which we have, instead of tending to increase it. This result has been so fully obtained, that most frequently this calculus has no real share in the complete solution of the question, except by its most simple parts; those which have reference to equations of the two first degrees, with one or more variables.

NUMERICAL RESOLUTION OF EQUATIONS.

The extreme imperfection of algebra, with respect to the resolution of equations, has led analysts to occupy themselves with a new class of questions, whose true character should be here noted. They have busied themselves in filling up the immense gap in the resolution of algebraic equations of the higher degrees, by what they have named the *numerical resolution* of equations. Not being able to obtain, in general, the *formula* which expresses what explicit function of the given quantities the unknown one is, they have sought (in the absence of this kind of resolution, the only one really *algebraic*) to determine, independently of that formula, at least the *value* of each unknown quantity, for various designated systems of particular values attributed to the given quantities. By the successive labours of analysts, this incomplete and illegitimate operation, which presents an intimate mixture of truly algebraic questions with others which are purely arithmetical, has been rendered possible in all cases for equations of any degree and even of any form. The methods for this which we now possess are sufficiently general, although the calculations to which they lead are often so complicated as to render it almost impossible to execute them. We have nothing else to do, then, in this part of algebra, but to simplify the methods sufficiently to render them regularly applicable, which we may hope hereafter to effect. In this condition of the calculus of direct functions, we endeavour, in its application, so to dispose the proposed questions as finally to require only this *numerical* resolution of the equations.

Its limited Usefulness. Valuable as is such a resource in the absence of the veritable solution, it is essential not to misconceive the true character of these methods, which analysts rightly regard as a very imperfect algebra. In fact, we are far from being always able to reduce our mathematical questions to depend finally upon only the *numerical* resolution of equations; that can be done only for questions quite isolated or truly final, that is, for the smallest number. Most questions, in fact, are only preparatory, and intended to serve as an indispensable preparation for the solution of other questions. Now, for such an object, it is evident that it is not the actual *value* of the unknown quantity which it is important to discover, but the *formula*, which shows how it is derived from the other quantities under consideration. It is this which happens, for example, in a very extensive class of cases, whenever a certain question includes at the same time several unknown quantities. We have then, first of all, to separate them. By suitably employing the simple and general method so happily invented by analysts, and which consists in referring all the other unknown quantities to one of them, the difficulty would always disappear if we knew how to obtain the algebraic resolution of the equations under consideration, while the *numerical* solution would then be perfectly useless. It is only for want of knowing the *algebraic* resolution of equations with a single unknown quantity, that we are obliged to treat *Elimination* as a distinct question, which forms one of the greatest special difficulties of common algebra. Laborious as are the methods by the aid of which we overcome this difficulty, they are not even applicable, in an entirely general manner, to the elimination of one unknown quantity between two equations of any form whatever.

In the most simple questions, and when we have really to resolve only a single equation with a single unknown quantity, this *numerical* resolution is none the less a very imperfect method, even when it is strictly sufficient. It presents, in fact, this serious inconvenience of obliging us to repeat the whole series of operations for the slightest change which may take place in a single one of the quantities considered, although their relations to one another remain unchanged; the calculations made for one case not enabling us to dispense with any of those which relate to a case very slightly different. This happens because of our inability to abstract and treat separately that purely algebraic part of the question which is common to all the cases which result from the mere variation of the given numbers.

According to the preceding considerations, the calculus of direct functions, viewed in its present state, divides into two very distinct branches, according as its subject is the *algebraic* resolution of equations or their *numerical* resolution. The first department, the only one truly satisfactory, is unhappily very limited, and will probably always remain so; the second, too often insufficient, has, at least, the advantage of a much greater generality. The necessity of clearly distinguishing these two parts is evident, because of the essentially different object proposed in each, and consequently the peculiar point of view under which quantities are therein considered.

Different Divisions of the two Methods of Resolution. If, moreover, we consider these parts with reference to the different methods of which each is composed, we find in their logical distribution an entirely different arrangement. In fact, the first part must be divided according to the nature of the equations which we are able to resolve, and independently of every consideration relative to the *values* of the unknown quantities. In the second part, on the contrary, it is not according to the *degrees* of the equations that the methods are naturally distinguished, since they are applicable to equations of any degree whatever; it is according to the numerical character of the *values* of the unknown quantities; for, in calculating these numbers directly, without deducing them from general formulas, different means would evidently be employed when the numbers are not susceptible of having their values determined otherwise than by a series of approximations, always incomplete, or when they can be obtained with entire exactness. This distinction of *incommensurable* and of *commensurable* roots, which require quite different principles for their determination, important as it is in the numerical resolution of equations, is entirely insignificant in the algebraic resolution, in which the *rational* or *irrational* nature of the numbers which are obtained is a mere accident of the calculation, which cannot exercise any influence over the methods employed; it is, in a word, a simple arithmetical consideration. We may say as much, though in a less degree, of the division of the commensurable roots themselves into *entire* and *fractional.* In fine, the case is the same, in a still greater degree, with the most general classification of roots, as *real* and *imaginary.* All these different considerations, which are preponderant as to the numerical resolution of equations, and which are of no importance in their algebraic resolution, render more and more sensible the essentially distinct nature of these two principal parts of algebra.

THE THEORY OF EQUATIONS.

These two departments, which constitute the immediate object of the calculus of direct functions, are subordinate to a third one, purely speculative, from which both of them borrow their most powerful resources, and which has been very exactly designated by the general name of *Theory of Equations*, although it as yet relates only to *Algebraic* equations. The numerical resolution of equations, because of its generality, has special need of this rational foundation.

This last and important branch of algebra is naturally divided into two orders of questions, viz., those which refer to the *composition* of equations, and those which concern their *transformation;* these latter having for their object to modify the roots of an equation without knowing them, in accordance with any given law, providing that this law is uniform in relation to all the parts.*

* The fundamental principle on which reposes the theory of equations, and which is so frequently applied in all mathematical analysis—the decomposition of algebraic, rational, and entire functions, of any degree whatever, into factors of the first degree—is never employed except for functions of a single variable, without any one having examined if it ought to be extended to functions of several variables. The general impossibility of such a decomposition is demonstrated by the author in detail, but more properly belongs to a special treatise.

THE METHOD OF INDETERMINATE COEFFICIENTS.

To complete this rapid general enumeration of the different essential parts of the calculus of direct functions, I must, lastly, mention expressly one of the most fruitful and important theories of algebra proper, that relating to the transformation of functions into series by the aid of what is called the *Method of indeterminate Coefficients.* This method, so eminently analytical, and which must be regarded as one of the most remarkable discoveries of Descartes, has undoubtedly lost some of its importance since the invention and the development of the infinitesimal calculus, the place of which it might so happily take in some particular respects. But the increasing extension of the transcendental analysis, although it has rendered this method much less necessary, has, on the other hand, multiplied its applications and enlarged its resources; so that by the useful combination between the two theories, which has finally been effected, the use of the method of indeterminate coefficients has become at present much more extensive than it was even before the formation of the calculus of indirect functions.

Having thus sketched the general outlines of algebra proper, I have now to offer some considerations on several leading points in the calculus of direct functions, our ideas of which may be advantageously made more clear by a philosophical examination.

IMAGINARY QUANTITIES.

The difficulties connected with several peculiar symbols to which algebraic calculations sometimes lead, and especially to the expressions called *imaginary*, have been, I think, much exaggerated through purely metaphysical considerations, which have been forced upon them, in the place of regarding these abnormal results in their true point of view as simple analytical facts. Viewing them thus, we readily see that, since the spirit of mathematical analysis consists in considering magnitudes in reference to their relations only, and without any regard to their determinate value, analysts are obliged to admit indifferently every kind of expression which can be engendered by algebraic combinations. The interdiction of even one expression because of its apparent singularity would destroy the generality of their conceptions. The common embarrassment on this subject seems to me to proceed essentially from an unconscious confusion between the idea of *function* and the idea of *value*, or, what comes to the same thing, between the *algebraic* and the *arithmetical* point of view. A thorough examination would show mathematical analysis to be much more clear in its nature than even mathematicians commonly suppose.

NEGATIVE QUANTITIES.

As to negative quantities, which have given rise to so many misplaced discussions, as irrational as useless, we must distinguish between their *abstract* signification and their *concrete* interpretation, which have been almost always confounded up to the present day. Under the first point of view, the theory of negative quantities can be established in a complete manner by a single algebraical consideration. The necessity of admitting such expressions is the same as for imaginary quantities, as above indicated; and their employment as an analytical artifice, to render the formulas more comprehensive, is a mechanism of calculation which cannot really give rise to any serious difficulty. We may therefore regard the abstract theory of negative quantities as leaving nothing essential to desire; it presents no obstacles but those inappropriately introduced by sophistical considerations.

It is far from being so, however, with their concrete theory. This consists essentially in that admirable property of the signs + and –, of representing analytically the oppositions of directions of which certain magnitudes are susceptible. This *general theorem* on the relation of the concrete to the abstract in mathematics is one of the most beautiful discoveries which we owe to the genius of Descartes, who obtained it as a simple result of properly directed philosophical observation. A great number of geometers have since striven to establish directly its general demonstration, but thus far their efforts have been illusory. Their vain metaphysical considerations and heterogeneous minglings of the abstract and the concrete have so confused the subject, that it becomes necessary to here distinctly enunciate the general fact. It consists in this: if, in any equation whatever, expressing the relation of certain quantities which are susceptible of opposition of directions, one or more of those quantities come to be reckoned in a direction contrary to that which belonged to them when the equation was first established, it will not be necessary to form directly a new equation for this second state of the phenomena; it will suffice to change, in the first equation, the sign of each of the quantities which shall have changed its direction; and the equation, thus modified, will always rigorously coincide with that which we would have arrived at in recommencing to investigate, for this new case, the analytical law of the phenomenon. The general theorem consists in this constant and necessary coincidence. Now, as yet, no one has succeeded in directly proving this; we have assured ourselves of it only by a great number of geometrical and mechanical verifications, which are, it is true, sufficiently multiplied, and especially sufficiently varied, to prevent any clear mind from having the least doubt of the exactitude and the generality of this essential property, but which, in a philosophical point of view, do not at all dispense with the research for so important an explanation. The extreme extent of the theorem must make us comprehend both the fundamental difficulties of this research and the high utility for the perfecting of mathematical science which would belong to the general conception of this great truth. This imperfection of theory, however, has not prevented geometers from making the most extensive and the most important use of this property in all parts of concrete mathematics.

It follows from the above general enunciation of the fact, independently of any demonstration, that the property of which we speak must never be applied to magnitudes whose directions are continually varying, without giving rise to a simple opposition of direction; in that case, the sign with which every result of calculation is necessarily affected is not susceptible of any concrete interpretation, and the attempts sometimes made to establish one are erroneous. This circumstance occurs, among other occasions, in the case of a radius vector in geometry, and diverging forces in mechanics.

PRINCIPLE OF HOMOGENEITY.

A second general theorem on the relation of the concrete to the abstract is that which is ordinarily designated under the name of *Principle of Homogeneity.* It is undoubtedly much less important in its applications than the preceding, but it particularly merits our attention as having, by its nature, a still greater extent, since it is applicable to all phenomena without distinction, and because of the real utility which it often possesses for the verification of their analytical laws. I can, moreover, exhibit a direct and general demonstration of it which seems to me very simple. It is founded on this single observation, which is self-evident, that the exactitude of every relation between any concrete magnitudes whatsoever is independent of the value of the *units* to which they are referred for the purpose of expressing them in numbers. For example, the relation which exists between the three sides of a right-angled triangle is the same, whether they are measured by yards, or by miles, or by inches.

It follows from this general consideration, that every equation which expresses the analytical law of any phenomenon must possess this property of being in no way altered, when all the quantities which are found in it are made to undergo simultaneously the change corresponding to that which their respective units would experience. Now this change evidently consists in all the quantities of each sort becoming at once m times smaller, if the unit which corresponds to them becomes m times greater, or reciprocally. Thus every equation which represents any concrete relation whatever must possess this characteristic of remaining the same, when we make m times greater all the quantities which it contains, and which express the magnitudes between which the relation exists; excepting always the numbers which designate simply the mutual *ratios* of these different magnitudes, and which therefore remain invariable during the change of the units. It is this property which constitutes the law of Homogeneity in its most extended signification, that is, of whatever analytical functions the equations may be composed.

But most frequently we consider only the cases in which the functions are such as are called *algebraic*, and to which the idea of *degree* is applicable. In this case we can give more precision to the general proposition by determining the analytical character which must be necessarily presented by the equation, in order that this property may be verified. It is easy to see, then, that, by the modification just explained, all the *terms* of the first degree, whatever may be their form, rational or irrational, entire or fractional, will become m times greater; all those of the second degree, m^2 times; those of the third, m^3 times, &c. Thus the terms of the same degree, however different may be their composition, varying in the same manner, and the terms of different degrees varying in an unequal proportion, whatever similarity there may be in their composition, it will be necessary, to prevent the equation from being disturbed, that all the terms which it contains should be of the same degree. It is in this that properly consists the ordinary theorem of *Homogeneity*, and it is from this circumstance that the general law has derived its name, which, however, ceases to be exactly proper for all other functions.

In order to treat this subject in its whole extent, it is important to observe an essential condition, to which attention must be paid in applying this property when the phenomenon expressed by the equation presents magnitudes of different natures. Thus it may happen that the respective units are completely independent of each other, and then the theorem of Homogeneity will hold good, either with reference to all the corresponding classes of quantities, or with regard to only a single one or more of them. But it will happen on other occasions that the different units will have fixed relations to one another, determined by the nature of the question; then it will be necessary to pay attention to this subordination of the units in verifying the homogeneity, which will not exist any longer in a purely algebraic sense, and the precise form of which will vary according to the nature of the phenomena. Thus, for example, to fix our ideas, when, in the analytical expression of geometrical phenomena, we are considering at once lines, areas, and volumes, it will be necessary to observe that the three corresponding units are necessarily so connected with each other that, according to the subordination generally established in that respect, when the first becomes m times greater, the second becomes m^2 times, and the third m^3 times. It is with such a modification that homogeneity will exist in the equations, in which, if they are *algebraic*, we will have to estimate the degree of each term by doubling the exponents of the factors which correspond to areas, and tripling those of the factors relating to volumes.

Such are the principal general considerations relating to the *Calculus of Direct Functions.* We have now to pass to the philosophical examination of the *Calculus of Indirect Functions*, the much superior importance and extent of which claim a fuller development.

CHAPTER III.

TRANSCENDENTAL ANALYSIS:

DIFFERENT MODES OF VIEWING IT.

WE determined, in the second chapter, the philosophical character of the transcendental analysis, in whatever manner it may be conceived, considering only the general nature of its actual destination as a part of mathematical science. This analysis has been presented by geometers under several points of view, really distinct, although necessarily equivalent, and leading always to identical results. They may be reduced to three principal ones; those of LEIBNITZ, of NEWTON, and of LAGRANGE, of which all the others are only secondary modifications. In the present state of science, each of these three general conceptions offers essential advantages which pertain to it exclusively, without our having yet succeeded in constructing a single method uniting all these different characteristic qualities. This combination will probably be hereafter effected by some method founded upon the conception of Lagrange. When that important philosophical labour shall have been accomplished, the study of the other conceptions will have only a historic interest; but, until then, the science must be considered as in only a provisional state, which requires the simultaneous consideration of all the various modes of viewing this calculus. Illogical as may appear this multiplicity of conceptions of one identical subject, still, without them all, we could form but a very insufficient idea of this analysis, whether in itself, or more especially in relation to its applications. This want of system in the most important part of mathematical analysis will not appear strange if we consider, on the one hand, its great extent and its superior difficulty, and, on the other, its recent formation.

ITS EARLY HISTORY.

If we had to trace here the systematic history of the successive formation of the transcendental analysis, it would be necessary previously to distinguish carefully from the *calculus of indirect functions*, properly so called, the original idea of the *infinitesimal method*, which can be conceived by itself, independently of any *calculus*. We should see that the first germ of this idea is found in the procedure constantly employed by the Greek geometers, under the name of the *Method of Exhaustions*, as a means of passing from the properties of straight lines to those of curves, and consisting essentially in substituting for the curve the auxiliary consideration of an inscribed or circumscribed polygon, by means of which they rose to the curve itself, taking in a suitable manner the limits of the primitive ratios. Incontestable as is this filiation of ideas, it would be giving it a greatly exaggerated importance to see in this method of exhaustions the real equivalent of our modern methods, as some geometers have done; for the ancients had no logical and general means for the determination of these limits, and this was commonly the greatest difficulty of the question; so that their solutions were not subjected to abstract and invariable rules, the uniform application of which would lead with certainty to the knowledge sought;

which is, on the contrary, the principal characteristic of our transcendental analysis. In a word, there still remained the task of generalizing the conceptions used by the ancients, and, more especially, by considering it in a manner purely abstract, of reducing it to a complete system of calculation, which to them was impossible.

The first idea which was produced in this new direction goes back to the great geometer Fermat, whom Lagrange has justly presented as having blocked out the direct formation of the transcendental analysis by his method for the determination of *maxima* and *minima*, and for the finding of *tangents*, which consisted essentially in introducing the auxiliary consideration of the correlative increments of the proposed variables, increments afterward suppressed as equal to zero when the equations had undergone certain suitable transformations. But, although Fermat was the first to conceive this analysis in a truly abstract manner, it was yet far from being regularly formed into a general and distinct calculus having its own notation, and especially freed from the superfluous consideration of terms which, in the analysis of Fermat, were finally not taken into the account, after having nevertheless greatly complicated all the operations by their presence. This is what Leibnitz so happily executed, half a century later, after some intermediate modifications of the ideas of Fermat introduced by Wallis, and still more by Barrow; and he has thus been the true creator of the transcendental analysis, such as we now employ it. This admirable discovery was so ripe (like all the great conceptions of the human intellect at the moment of their manifestation), that Newton, on his side, had arrived, at the same time, or a little earlier, at a method exactly equivalent, by considering this analysis under a very different point of view, which, although more logical in itself, is really less adapted to give to the common fundamental method all the extent and the facility which have been imparted to it by the ideas of Leibnitz. Finally, Lagrange, putting aside the heterogeneous considerations which had guided Leibnitz and Newton, has succeeded in reducing the transcendental analysis, in its greatest perfection, to a purely algebraic system, which only wants more aptitude for its practical applications.

After this summary glance at the general history of the transcendental analysis, we will proceed to the dogmatic exposition of the three principal conceptions, in order to appreciate exactly their characteristic properties, and to show the necessary identity of the methods which are thence derived. Let us begin with that of Leibnitz.

METHOD OF LEIBNITZ.

Infinitely small Elements. This consists in introducing into the calculus, in order to facilitate the establishment of equations, the infinitely small elements of which all the quantities, the relations between which are sought, are considered to be composed. These elements or *differentials* will have certain relations to one another, which are constantly and necessarily more simple and easy to discover than those of the primitive quantities, and by means of which we will be enabled (by a special calculus having for its peculiar object the elimination of these auxiliary infinitesimals) to go back to the desired equations, which it would have been most frequently impossible to obtain directly. This indirect analysis may have different degrees of indirectness; for, when there is too much difficulty in forming immediately the equation between the differentials of the magnitudes under consideration, a second application of the same general artifice will have to be made, and these differentials be treated, in their turn, as new primitive quantities, and a relation be sought between their infinitely small elements (which, with reference to the final objects of the question, will be *second differentials*), and so on; the same transformation admitting of being repeated any number of times, on the condition of finally eliminating the constantly increasing number of infinitesimal quantities introduced as auxiliaries.

A person not yet familiar with these considerations does not perceive at once how the employment of these auxiliary quantities can facilitate the discovery of the analytical laws of phenomena; for the infinitely small increments of the proposed magnitudes being of the same species with them, it would seem that their relations should not be obtained with more ease, inasmuch as the greater or less value of a quantity cannot, in fact, exercise any influence on an inquiry which is necessarily independent, by its nature, of every idea of value. But it is easy, nevertheless, to explain very clearly, and in a quite general manner, how far the question must be simplified by such an artifice. For this purpose, it is necessary to begin by distinguishing *different orders* of infinitely small quantities, a very precise idea of which may be obtained by considering them as being either the successive powers of the same primitive infinitely small quantity, or as being quantities which may be regarded as having finite ratios with these powers; so that, to take an example, the second, third, &c., differentials of any one variable are classed as infinitely small quantities of the second order, the third, &c., because it is easy to discover in them finite multiples of the second, third, &c., powers of a certain first differential. These preliminary ideas being established, the spirit of the infinitesimal analysis consists in constantly neglecting the infinitely small quantities in comparison with finite quantities, and generally the infinitely small quantities of any order whatever in comparison with all those of an inferior order. It is at once apparent how much such a liberty must facilitate the formation of equations between the differentials of quantities, since, in the place of these differentials, we can substitute such other elements as we may choose, and as will be more simple to consider, only taking care to conform to this single condition, that the new elements differ from the preceding ones only by quantities infinitely small in comparison with them. It is thus that it will be possible, in geometry, to treat curved lines as composed of an infinity of rectilinear elements, curved surfaces as formed of plane elements, and, in mechanics, variable motions as an infinite series of uniform motions, succeeding one another at infinitely small intervals of time.

Examples. Considering the importance of this admirable conception, I think that I ought here to complete the illustration of its fundamental character by the summary indication of some leading examples.

1. *Tangents.* Let it be required to determine, for each point of a plane curve, the equation of which is given, the direction of its tangent; a question whose general solution was the primitive object of the invent-

ors of the transcendental analysis. We will consider th tangent as a secant joining two points infinitely near to each other; and then, designating by dy and dx the infinitely small differences of the co-ordinates of those two points, the elementary principles of geometry will immediately give the equation $t=\frac{dy}{dx}$ for the trigonometrical tangent of the angle which is made with the axis of the abscissas by the desired tangent, this being the most simple way of fixing its position in a system of rectilinear co-ordinates. This equation, common to all curves, being established, the question is reduced to a simple analytical problem, which will consist in eliminating the infinitesimals dx and dy, which were introduced as auxiliaries, by determining in each particular case, by means of the equation of the proposed curve, the ratio of dy to dx, which will be constantly done by uniform and very simple methods.

2. *Rectification of an Arc.* In the second place, suppose that we wish to know the length of the arc of any curve, considered as a function of the co-ordinates of its extremities. It would be impossible to establish directly the equation between this arc s and these co-ordinates, while it is easy to find the corresponding relation between the differentials of these different magnitudes. The most simple theorems of elementary geometry will in fact give at once, considering the infinitely small arc ds as a right line, the equations

$$ds^2=dy^2+dx^2, \text{ or } ds^2=dx^2+dy^2+dz^2,$$

according as the curve is of single or double curvature. In either case, the question is now entirely within the domain of analysis, which, by the elimination of the differentials (which is the peculiar object of the calculus of indirect functions), will carry us back from this relation to that which exists between the finite quantities themselves under examination.

3. *Quadrature of a Curve.* It would be the same with the quadrature of curvilinear areas. If the curve is a plane one, and referred to rectilinear co-ordinates, we will conceive the area A comprised between this curve, the axis of the abscissas, and two extreme co-ordinates, to increase by an infinitely small quantity dA, as the result of a corresponding increment of the abscissa. The relation between these two differentials can be immediately obtained with the greatest facility by substituting for the curvilinear element of the proposed area the rectangle formed by the extreme ordinate and the element of the abscissa, from which it evidently differs only by an infinitely small quantity of the second order. This will at once give, whatever may be the curve, the very simple differential equation

$$dA=ydx,$$

from which, when the curve is defined, the calculus of indirect functions will show how to deduce the finite equation, which is the immediate object of the problem.

4. *Velocity in Variable Motion.* In like manner, in Dynamics, when we desire to know the expression for the velocity acquired at each instant by a body impressed with a motion varying according to any law, we will consider the motion as being uniform during an infinitely small element of the time t, and we will thus immediately form the differential equation $de=vdt$, in which v designates the velocity acquired when the body has passed over the space e; and thence it will be easy to deduce, by simple and invariable analytical procedures,

the formula which would give the velocity in each particular motion, in accordance with the corresponding relation between the time and the space; or, reciprocally, what this relation would be if the mode of variation of the velocity was supposed to be known, whether with respect to the space or to the time.

5. *Distribution of Heat.* Lastly, to indicate another kind of questions, it is by similar steps that we are able, in the study of thermological phenomena, according to the happy conception of M. Fourier, to form in a very simple manner the general differential equation which expresses the variable distribution of heat in any body whatever, subjected to any influences, by means of the single and easily-obtained relation, which represents the uniform distribution of heat in a right-angled parallelopipedon, considering (geometrically) every other body as decomposed into infinitely small elements of a similar form, and (thermologically) the flow of heat as constant during an infinitely small element of time. Henceforth, all the questions which can be presented by abstract thermology will be reduced, as in geometry and mechanics, to mere difficulties of analysis, which will always consist in the elimination of the differentials introduced as auxiliaries to facilitate the establishment of the equations.

Examples of such different natures are more than sufficient to give a clear general idea of the immense scope of the fundamental conception of the transcendental analysis as formed by Leibnitz, constituting, as it undoubtedly does, the most lofty thought to which the human mind has as yet attained.

It is evident that this conception was indispensable to complete the foundation of mathematical science, by enabling us to establish, in a broad and fruitful manner, the relation of the concrete to the abstract. In this respect it must be regarded as the necessary complement of the great fundamental idea of Descartes on the general analytical representation of natural phenomena: an idea which did not begin to be worthily appreciated and suitably employed till after the formation of the infinitesimal analysis, without which it could not produce, even in geometry, very important results.

Generality of the Formulas. Besides the admirable facility which is given by the transcendental analysis for the investigation of the mathematical laws of all phenomena, a second fundamental and inherent property, perhaps as important as the first, is the extreme generality of the differential formulas, which express in a single equation each determinate phenomenon, however varied the subjects in relation to which it is considered. Thus we see, in the preceding examples, that a single differential equation gives the tangents of all curves, another their rectifications, a third their quadratures; and in the same way, one invariable formula expresses the mathematical law of every variable motion; and, finally, a single equation constantly represents the distribution of heat in any body and for any case. This generality, which is so exceedingly remarkable, and which is for geometers the basis of the most elevated considerations, is a fortunate and necessary consequence of the very spirit of the transcendental analysis, especially in the conception of Leibnitz. Thus the infinitesimal analysis has not only furnished a general method for indirectly forming equations which it would have been impossible to discover in a direct manner, but it has also permitted us to consider, for the mathematical study of natural phenomena, a new order of more general laws, which nevertheless present a clear and precise signification to every mind habituated to their interpretation. By virtue of this second characteristic property, the entire system of an immense science, such as geometry or mechanics, has been condensed into a small number of analytical formulas, from which the human mind can deduce, by certain and invariable rules, the solution of all particular problems.

Demonstration of the Method. To complete the general exposition of the conception of Leibnitz, there remains to be considered the demonstration of the logical procedure to which it leads, and this, unfortunately, is the most imperfect part of this beautiful method.

In the beginning of the infinitesimal analysis, the most celebrated geometers rightly attached more importance to extending the immortal discovery of Leibnitz and multiplying its applications than to rigorously establishing the logical bases of its operations. They contented themselves for a long time by answering the objections of second-rate geometers by the unhoped-for solution of the most difficult problems; doubtless persuaded that in mathematical science, much more than in any other, we may boldly welcome new methods, even when their rational explanation is imperfect, provided they are fruitful in results, inasmuch as its much easier and more numerous verifications would not permit any error to remain long undiscovered. But this state of things could not long exist, and it was necessary to go back to the very foundations of the analysis of Leibnitz in order to prove, in a perfectly general manner, the rigorous exactitude of the procedures employed in this method, in spite of the apparent infractions of the ordinary rules of reasoning which it permitted.

Leibnitz, urged to answer, had presented an explanation entirely erroneous, saying that he treated infinitely small quantities as *incomparables*, and that he neglected them in comparison with finite quantities, "like grains of sand in comparison with the sea:" a view which would have completely changed the nature of his analysis, by reducing it to a mere approximative calculus, which, under this point of view, would be radically vicious, since it would be impossible to foresee, in general, to what degree the successive operations might increase these first errors, which could thus evidently attain any amount. Leibnitz, then, did not see, except in a very confused manner, the true logical foundations of the analysis which he had created. His earliest successors limited themselves, at first, to verifying its exactitude by showing the conformity of its results, in particular applications, to those obtained by ordinary algebra or the geometry of the ancients; reproducing, according to the ancient methods, so far as they were able, the solutions of some problems after they had been once obtained by the new method, which alone was capable of discovering them in the first place.

When this great question was considered in a more general manner, geometers, instead of directly attacking the difficulty, preferred to elude it in some way, as Euler and D'Alembert, for example, have done, by demonstrating the necessary and constant conformity of the conception of Leibnitz, viewed in all its applications, with other fundamental conceptions of the transcendental analysis, that of Newton especially, the exactitude of which was free from any objection. Such a general verification is undoubtedly strictly sufficient to dissipate any uncertainty as to the legitimate employment of the analysis of Leibnitz. But the infinitesimal method is so important—it offers still, in almost all its applications, such a practical superiority over the other general conceptions which have been successively proposed—that there would be a real imperfection in the philosophical character of the science if it could not justify itself, and needed to be logically founded on considerations of another order, which would then cease to be employed.

It was, then, of real importance to establish directly and in a general manner the necessary rationality of the infinitesimal method. After various attempts more or less imperfect, a distinguished geometer, Carnot, presented at last the true direct logical explanation of the method of Leibnitz, by showing it to be founded on the principle of the necessary compensation of errors, this being, in fact, the precise and luminous manifestation of what Leibnitz had vaguely and confusedly perceived. Carnot has thus rendered the science an essential service, although, as we shall see towards the end of this chapter, all this logical scaffolding of the infinitesimal method, properly so called, is very probably susceptible of only a provisional existence, inasmuch as it is radically vicious in its nature. Still, we should not fail to notice the general system of reasoning proposed by Carnot, in order to directly legitimate the analysis of Leibnitz. Here is the substance of it:

In establishing the differential equation of a phenomenon, we substitute, for the immediate elements of the different quantities considered, other simpler infinitesimals, which differ from them infinitely little in comparison with them; and this substitution constitutes the principal artifice of the method of Leibnitz, which without it would possess no real facility for the formation of equations. Carnot regards such an hypothesis as really producing an error in the equation thus obtained, and which for this reason he calls *imperfect;* only, it is clear that this error must be infinitely small. Now, on the other hand, all the analytical operations, whether of differentiation or of integration, which are performed upon these differential equations, in order to raise them to finite equations by eliminating all the infinitesimals which have been introduced as auxiliaries, produce as constantly, by their nature, as is easily seen, other analogous errors, so that an exact compensation takes place, and the final equations, in the words of Carnot, become *perfect.* Carnot views, as a certain and invariable indication of the actual establishment of this necessary compensation, the complete elimination of the various infinitely small quantities, which is always, in fact, the final object of all the operations of the transcendental analysis; for if we have committed no other infractions of the general rules of reasoning than those thus exacted by the very nature of the infinitesimal method, the infinitely small errors thus produced cannot have engendered other than infinitely small errors in all the equations, and the relations are necessarily of a rigorous exactitude as soon as they exist between finite quantities alone, since the only errors then possible must be finite ones, while none such can have entered. All this general reasoning is founded on the conception of infinitesimal quantities, regarded as indefinitely decreasing, while those from which they are derived are regarded as fixed.

Illustration by Tangents. Thus, to illustrate this abstract exposition by a single example, let us take up again the question of *tangents*, which is the most easy to analyze completely. We will regard the equation $t=\frac{dy}{dx}$, obtained above, as being affected with an infinitely small error, since it would be perfectly rigorous only for the secant. Now let us complete the solution by seeking, according to the equation of each curve, the ratio between the differentials of the co-ordinates. If we suppose this equation to be $y=ax^2$, we shall evidently have

$$dy=2axdx+adx^2.$$

In this formula we shall have to neglect the term dx^2 as an infinitely small quantity of the second order. Then the combination of the two *imperfect* equations.

$$t=\frac{dy}{dx},\ dy=2axdx,$$

being sufficient to eliminate entirely the infinitesimals, the finite result, $t=2ax$, will necessarily be rigorously correct, from the effect of the exact compensation of the two errors committed ; since, by its finite nature, it cannot be affected by an infinitely small error, and this is, nevertheless, the only one which it could have, according to the spirit of the operations which have been executed.

It would be easy to reproduce in a uniform manner the same reasoning with reference to all the other general applications of the analysis of Leibnitz.

This ingenious theory is undoubtedly more subtile than solid, when we examine it more profoundly ; but it has really no other radical logical fault than that of the infinitesimal method itself, of which it is, it seems to me, the natural development and the general explanation, so that it must be adopted for as long a time as it shall be thought proper to employ this method directly.

I pass now to the general exposition of the two other fundamental conceptions of the transcendental analysis, limiting myself in each to its principal idea, the philosophical character of the analysis having been sufficiently determined above in the examination of the conception of Leibnitz, which I have specially dwelt upon because it admits of being most easily grasped as a whole, and most rapidly described.

METHOD OF NEWTON.

Newton has successively presented his own method of conceiving the transcendental analysis under several different forms. That which is at present the most commonly adopted was designated by Newton, sometimes under the name of the *Method of prime and ultimate Ratios*, sometimes under that of the *Method of Limits.*

Method of Limits. The general spirit of the transcendental analysis, from this point of view, consists in introducing as auxiliaries, in the place of the primitive quantities, or concurrently with them, in order to facilitate the establishment of equations, the *limits of the ratios* of the simultaneous increments of these quantities; or, in other words, the *final ratios* of these increments; limits or final ratios which can be easily shown to have a determinate and finite value. A special calculus, which is the equivalent of the infinitesimal calculus, is then employed to pass from the equations between these limits to the corresponding equations between the primitive quantities themselves.

The power which is given by such an analysis, of expressing with more ease the mathematical laws of phenomena, depends in general on this, that since the calculus applies, not to the increments themselves of the proposed quantities, but to the limits of the ratios of those increments, we can always substitute for each increment any other magnitude more easy to consider, provided that their final ratio is the ratio of equality, or, in other words, that the limit of their ratio is unity. It is clear, indeed, that the calculus of limits would be in no way affected by this substitution. Starting from this principle, we find nearly the equivalent of the facilities offered by the analysis of Leibnitz, which are then merely conceived under another point of view. Thus curves will be regarded as the *limits* of a series of rectilinear polygons, variable motions as the *limits* of a collection of uniform motions of constantly diminishing durations, and so on.

EXAMPLES. 1. *Tangents.* Suppose, for example, that we wish to determine the direction of the tangent to a curve; we will regard it as the limit towards which would tend a secant, which should turn about the given point so that its second point of intersection should indefinitely approach the first. Representing the differences of the coordinates of the two points by $\triangle y$ and $\triangle x$, we would have at each instant, for the trigonometrical tangent of the angle which the secant makes with the axis of abscissas,

$$t=\frac{\triangle y}{\triangle x};$$

from which, taking the limits, we will obtain, relatively to the tangent itself, this general formula of transcendental analysis,

$$t=L\frac{\triangle y}{\triangle x},$$

the characteristic L being employed to designate the limit. The calculus of indirect functions will show how to deduce from this formula in each particular case, when the equation of the curve is given, the relation between t and x, by eliminating the auxiliary quantities which have been introduced. If we suppose, in order to complete the solution, that the equation of the proposed curve is $y=ax^2$, we shall evidently have

$$\triangle y=2ax\triangle x+a(\triangle x)^2,$$

from which we shall obtain

$$\frac{\triangle y}{\triangle x}=2ax+a\triangle x.$$

Now it is clear that the *limit* towards which the second number tends, in proportion as $\triangle x$ diminishes, is $2ax$. We shall therefore find, by this method, $t=2ax$, as we obtained it for the same case by the method of Leibnitz.

2. *Rectifications.* In like manner, when the rectification of a curve is desired, we must substitute for the increment of the arc s the chord of this increment, which evidently has such a connexion with it that the limit of their ratio is unity; and then we find (pursuing in other respects the same plan as with the method of Leibnitz) this general equation of rectifications:

$$\left(L\frac{\triangle s}{\triangle x}\right)^2=1+\left(L\frac{\triangle y}{\triangle x}\right)^2,$$

or

$$\left(L\frac{\triangle s}{\triangle x}\right)^2=1+\left(L\frac{\triangle y}{\triangle x}\right)^2+\left(L\frac{\triangle z}{\triangle x}\right)^2,$$

according as the curve is plane or of double curvature. It will now be necessary, for each particular curve, to pass from this equation to that between the arc and the abscissa, which depends on the transcendental calculus properly so called.

We could take up, with the same facility, by the method of limits, all the other general questions, the solution of which has been already indicated according to the infinitesimal method.

Such is, in substance, the conception which Newton formed for the transcendental analysis, or, more precisely, that which Maclaurin and D'Alembert have presented as the most rational basis of that analysis, in seeking to fix and to arrange the ideas of Newton upon that subject.

Fluxions and Fluents. Another distinct form under which Newton has presented this same method should be here noticed, and deserves particularly to fix our attention, as much by its ingenious clearness in some cases as by its having furnished the notation best suited to this manner of viewing the transcendental analysis, and, moreover, as having been till lately the special form of the calculus of indirect functions commonly adopted by the English geometers. I refer to the calculus of *fluxions* and of *fluents*, founded on the general idea of *velocities*.

To facilitate the conception of the fundamental idea, let us consider every curve as generated by a point impressed with a motion varying according to any law whatever. The different quantities which the curve can present, the abscissa, the ordinate, the arc, the area, &c., will be regarded as simultaneously produced by successive degrees during this motion. The *velocity* with which each shall have been described will be called the *fluxion* of that quantity, which will be inversely named its *fluent*. Henceforth the transcendental analysis will consist, according to this conception, in forming directly the equations between the fluxions of the proposed quantities, in order to deduce therefrom, by a special calculus, the equations between the fluents themselves. What has been stated respecting curves may, moreover, evidently be applied to any magnitudes whatever, regarded, by the aid of suitable images, as produced by motion.

It is easy to understand the general and necessary identity of this method with that of limits complicated with the foreign idea of motion. In fact, resuming the case of the curve, if we suppose, as we evidently always may, that the motion of the describing point is uniform in a certain direction, that of the abscissa, for example, then the fluxion of the abscissa will be constant, like the element of the time; for all the other quantities generated, the motion cannot be conceived to be uniform, except for an infinitely small time. Now the velocity being in general according to its mechanical conception, the ratio of each space to the time employed in traversing it, and this time being here proportional to the increment of the abscissa, it follows that the fluxions of the ordinate, of the arc, of the area, &c., are really nothing else (rejecting the intermediate consideration of time) than the final ratios of the increments of these different quantities to the increment of the abscissa. This method of fluxions and fluents is, then, in reality, only a manner of representing, by a comparison borrowed from mechanics, the method of prime and ultimate ratios, which alone can be reduced to a calculus. It evidently, then, offers the same general advantages in the various principal applications of the transcendental analysis, without its being necessary to present special proofs of this.

METHOD OF LAGRANGE.

Derived Functions. The conception of Lagrange, in its admirable simplicity, consists in representing the transcendental analysis as a great algebraic artifice, by which, in order to facilitate the establishment of equations, we introduce, in the place of the primitive functions, or concurrently with them, their *derived* functions; that is, according to the definition of Lagrange, the coefficient of the first term of the increment of each function, arranged according to the ascending powers of the increment of its variable. The special calculus of indirect functions has for its constant object, here as well as in the conceptions of Leibnitz and of Newton, to eliminate these *derivatives* which have been thus employed as auxiliaries, in order to deduce from their relations the corresponding equations between the primitive magnitudes.

An Extension of ordinary Analysis. The transcendental analysis is, then, nothing but a simple though very considerable extension of ordinary analysis. Geometers have long been accustomed to introduce in analytical investigations, in the place of the magnitudes themselves which they wished to study, their different powers, or their logarithms, or their sines, &c., in order to simplify the equations, and even to obtain them more easily. This successive *derivation* is an artifice of the same nature, only of greater extent, and procuring, in consequence, much more important resources for this common object.

But, although we can readily conceive, *à priori*, that the auxiliary consideration of these derivatives *may* facilitate the establishment of equations, it is not easy to explain why this *must* necessarily follow from this mode of derivation rather than from any other transformation. Such is the weak point of the great idea of Lagrange. The precise advantages of this analysis cannot as yet be grasped in an abstract manner, but only shown by considering separately each principal question, so that the verification is often exceedingly laborious.

EXAMPLE. *Tangents.* This manner of conceiving the transcendental analysis may be best illustrated by its application to the most simple of the problems above examined—that of tangents.

Instead of conceiving the tangent as the prolongation of the infinitely small element of the curve, according to the notion of Leibnitz—or as the limit of the secants, according to the ideas of Newton—Lagrange considers it, according to its simple geometrical character, analogous to the definitions of the ancients, to be a right line such that no other right line can pass through the point of contact between it and the curve. Then, to determine its direction, we must seek the general expression of its distance from the curve, measured in any direction whatever—in that of the ordinate, for example—and dispose of the arbitrary constant relating to the inclination of the right line, which will necessarily enter into that expression, in such a way as to diminish that separation as much as possible. Now this distance, being evidently equal to the difference of the two ordinates of the curve and of the right line, which correspond to the same new abscissa $x+h$, will be represented by the formula

$$\big(f'(x)-t\big)h+qh^2+rh^3+\text{ etc.},$$

in which t designates, as above, the unknown trigonometrical tangent of the angle which the required line makes with the axis of abscissas, and $f'(x)$ the derived function of the ordinate $f(x)$. This being understood, it is easy to see that, by disposing of t so as to make the first term of the preceding formula equal to zero, we will render the interval between the two lines the least possible, so that any other line for which t did not have the value thus determined would necessarily depart farther from the proposed curve. We have, then, for the direction of the tangent sought, the general expression $t=f'(x)$, a result exactly equivalent to those furnished by the Infinitesimal Method and the Method of Limits. We have yet to find $f'(x)$ in each particular curve, which is a mere question of analysis, quite identical with those which are presented, at this stage of the operations, by the other methods.

After these considerations upon the principal general conceptions, we need not stop to examine some other theories proposed, such as Euler's *Calculus of Vanishing Quantities*, which are really modifications—more or less important, and, moreover, no longer used—of the preceding methods.

I have now to establish the comparison and the appreciation of these three fundamental methods. Their *perfect and necessary conformity* is first to be proven in a general manner.

FUNDAMENTAL IDENTITY OF THE THREE METHODS.

It is, in the first place, evident from what precedes, considering these three methods as to their actual destination, independently of their preliminary ideas, that they all consist in the same general logical artifice, which has been characterized in the first chapter; to wit, the introduction of a certain system of auxiliary magnitudes, having uniform relations to those which are the special objects of the inquiry, and substituted for them expressly to facilitate the analytical expression of the mathematical laws of the phenomena, although they have finally to be eliminated by the aid of a special calculus. It is this which has determined me to regularly define the transcendental analysis as *the calculus of indirect functions*, in order to mark its true philosophical character, at the same time avoiding any discussion upon the best manner of conceiving and applying it. The general effect of this analysis, whatever the method employed, is, then, to bring every mathematical question much more promptly within the power of the *calculus*, and thus to diminish considerably the serious difficulty which is usually presented by the passage from the concrete to the abstract. Whatever progress we may make, we can never hope that the calculus will ever be able to grasp every question of natural philosophy, geometrical, or mechanical, or thermological, &c., immediately upon its birth, which would evidently involve a contradiction. Every problem will constantly require a certain preliminary labour to be performed, in which the calculus can be of no assistance, and which, by its nature, cannot be subjected to abstract and invariable rules; it is that which has for its special object the establishment of equations, which form the indispensable starting point of all analytical researches. But this preliminary labour has been remarkably simplified by the creation of the transcendental analysis, which has thus hastened the moment at which the solution admits of the uniform and precise application of general and abstract methods; by reducing, in each case, this special labour to the investigation of equations between the auxiliary magnitudes; from which the calculus then leads to equations directly referring to the proposed magnitudes, which, before this admirable conception, it had been necessary to establish directly and separately. Whether these indirect equations are *differential* equations, according to the idea of Leibnitz, or equations of *limits*, conformably to the conception of Newton, or, lastly, *derived* equations, according to the theory of Lagrange, the general procedure is evidently always the same.

But the coincidence of these three principal methods is not limited to the common effect which they produce; it exists, besides, in the very manner of obtaining it. In fact, not only do all three consider, in the place of the primitive magnitudes, certain auxiliary ones, but, still farther, the quantities thus introduced as subsidiary are exactly identical in the three methods, which consequently differ only in the manner of viewing them. This can be easily shown by taking for the general term of comparison any one of the three conceptions, especially that of Lagrange, which is the most suitable to serve as a type, as being the freest from foreign considerations. Is it not evident, by the very definition of *derived functions*, that they are nothing else than what Leibnitz calls *differential coefficients*, or the ratios of the differential of each function to that of the corresponding variable, since, in determining the first differential, we will be obliged, by the very nature of the infinitesimal method, to limit ourselves to taking the only term of the increment of the function which contains the first power of the infinitely small increment of the variable? In the same way, is not the derived function, by its nature, likewise the necessary *limit* towards which tends the ratio between the increment of the primitive function and that of its variable, in proportion as this last indefinitely diminishes, since it evidently expresses what that ratio becomes when we suppose the increment of the variable to equal zero? That which is designated by $\frac{dy}{dx}$ in the method of Leibnitz; that which ought to be noted as $L\frac{\triangle y}{\triangle x}$ in that of Newton; and that which Lagrange has indicated by $f'(x)$, is constantly one same function, seen from three different points of view, the considerations of Leibnitz and Newton properly consisting in making known two general necessary properties of the derived function. The transcendental analysis, examined abstractedly and in its principle, is then always the same, whatever may be the conception which is adopted, and the procedures of the calculus of indirect functions are necessarily identical in these different methods, which in like manner must, for any application whatever, lead constantly to rigorously uniform results.

COMPARATIVE VALUE OF THE THREE METHODS.

If now we endeavour to estimate the comparative value of these three equivalent conceptions, we shall find in each advantages and inconveniences which are peculiar to it, and which still prevent geometers from confining themselves to any one of them, considered as final.

That of Leibnitz. The conception of Leibnitz presents incontestably, in all its applications, a very marked superiority, by leading in a much more rapid manner, and with much less mental effort, to the formation of equations between the auxiliary magnitudes. It is to its use that we owe the high perfection which has been acquired by all the general theories of geometry and mechanics. Whatever may be the different speculative opinions of geometers with respect to the infinitesimal method, in an abstract point of view, all tacitly agree in employing it by preference, as soon as they have to treat a new question, in order not to complicate the necessary difficulty by this purely artificial obstacle proceeding from a misplaced obstinacy in adopting a less expeditious course. Lagrange himself, after having reconstructed the transcendental analysis on new foundations, has (with that noble frankness which so well suited his genius) rendered a striking and decisive homage to the characteristic properties of the conception of Leibnitz, by following it exclusively in the entire system of his *Mecanique Analytique*. Such a fact renders any comments unnecessary.

But when we consider the conception of Leibnitz in itself and in its logical relations, we cannot escape admitting, with Lagrange, that it is radically vicious in this, that, adopting its own expressions, the notion of infinitely small quantities is a *false idea*, of which it is in fact impossible to obtain a clear conception, however we may deceive ourselves in that matter. Even if we adopt the ingenious idea of the compensation of errors, as above explained, this involves the radical inconvenience of being obliged to distinguish in mathematics two classes of reasonings, those which are perfectly rigorous, and those in which we designedly commit errors which subsequently have to be compensated. A conception which leads to such strange consequences is undoubtedly very unsatisfactory in a logical point of view.

To say, as do some geometers, that it is possible in every case to reduce the infinitesimal method to that of limits, the logical character of which is irreproachable, would evidently be to elude the difficulty rather than to remove it; besides, such a transformation almost entirely strips the conception of Leibnitz of its essential advantages of facility and rapidity.

Finally, even disregarding the preceding important considerations, the infinitesimal method would no less evidently present by its nature the very serious defect of breaking the unity of abstract mathematics, by creating a transcendental analysis founded on principles so different from those which form the basis of the ordinary analysis. This division of analysis into two worlds almost entirely independent of each other, tends to hinder the formation of truly general analytical conceptions. To fully appreciate the consequences of this, we should have to go back to the state of the science before Lagrange had established a general and complete harmony between these two great sections.

That of Newton. Passing now to the conception of Newton, it is evident that by its nature it is not exposed to the fundamental logical objections which are called forth by the method of Leibnitz. The notion of *limits* is, in fact, remarkable for its simplicity and its precision. In the transcendental analysis presented in this manner, the equations are regarded as exact from their very origin, and the general rules of reasoning are as constantly observed as in ordinary analysis. But, on the other hand, it is very far from offering such powerful resources for the solution of problems as the infinitesimal method. The obligation which it imposes, of never considering the increments of magnitudes separately and by themselves, nor even in their ratios, but only in the limits of those ratios, retards considerably the operations of the mind in the formation of auxiliary equations. We may even say that it greatly embarrasses the purely analytical transformations. Thus the transcendental analysis, considered separately from its applications, is far from presenting in this method the extent and the generality which have been imprinted upon it by the conception of Leibnitz. It is very difficult, for example, to extend the theory of Newton to functions of several independent variables. But it is especially with reference to its applications that the relative inferiority of this theory is most strongly marked.

Several Continental geometers, in adopting the method of Newton as the more logical basis of the transcendental analysis, have partially disguised this inferiority by a serious inconsistency, which consists in applying to this method the notation invented by Leibnitz for the infinitesimal method, and which is really appropriate to it alone. In designating by $\frac{dy}{dx}$ that which logically ought, in the theory of limits, to be denoted by $L\frac{\triangle y}{\triangle x}$, and in extending to all the other analytical conceptions this displacement of signs, they intended, undoubtedly, to combine the special advantages of the two methods; but, in reality, they have only succeeded in causing a vicious confusion between them, a familiarity with which hinders the formation of clear and exact ideas of either. It would certainly be singular, considering this usage in itself, that, by the mere means of signs, it could be possible to effect a veritable combination between two theories so distinct as those under consideration.

Finally, the method of limits presents also, though in a less degree, the greater inconvenience, which I have above noted in reference to the infinitesimal method, of establishing a total separation between the ordinary and the transcendental analysis; for the idea of *limits*, though clear and rigorous, is none the less in itself, as Lagrange has remarked, a foreign idea, upon which analytical theories ought not to be dependent.

That of Lagrange. This perfect unity of analysis, and this purely abstract character of its fundamental notions, are found in the highest degree in the conception of Lagrange, and are found there alone; it is, for this reason, the most rational and the most philosophical of all. Carefully removing every heterogeneous consideration, Lagrange has reduced the transcendental analysis to its true peculiar character, that of presenting a very extensive class of analytical transformations, which facilitate in a remarkable degree the expression of the conditions of various problems. At the same time, this analysis is thus necessarily presented as a simple extension of ordinary analysis; it is only a higher algebra. All the different parts of abstract mathematics, previously so incoherent, have from that moment admitted of being conceived as forming a single system.

Unhappily, this conception, which possesses such fundamental properties, independently of its so simple and so lucid notation, and which is undoubtedly destined to become the final theory of transcendental analysis, because of its high philosophical superiority over all the other methods proposed, presents in its present state too many difficulties in its applications, as compared with the conception of Newton, and still more with that of Leibnitz, to be as yet exclusively adopted. Lagrange himself has succeeded only with great difficulty in rediscovering, by his method, the principal results already obtained by the infinitesimal method for the solution of the general questions of geometry and mechanics ; we may judge from that what obstacles would be found in treating in the same manner questions which were truly new and important. It is true that Lagrange, on several occasions, has shown that difficulties call forth, from men of genius, superior efforts, capable of leading to the greatest results. It was thus that, in trying to adapt his method to the examination of the curvature of lines, which seemed so far from admitting its application, he arrived at that beautiful theory of contacts which has so greatly perfected that important part of geometry. But, in spite of such happy exceptions, the conception of Lagrange has nevertheless remained, as a whole, essentially unsuited to applications.

The final result of the general comparison which I have too briefly sketched, is, then, as already suggested, that, in order to really understand the transcendental analysis, we should not only consider it in its principles according to the three fundamental conceptions of Leibnitz, of Newton, and of Lagrange, but should besides accustom ourselves to carry out almost indifferently, according to these three principal methods, and especially according to the first and the last, the solution of all important questions, whether of the pure calculus of indirect functions or of its applications. This is a course which I could not too strongly recommend to all those who desire to judge philosophically of this admirable creation of the human mind, as well as to those who wish to learn to make use of this powerful instrument with success and with facility. In all the other parts of mathematical science, the consideration of different methods for a single class of questions may be useful, even independently of its historical interest, but it is not indispensable; here, on the contrary, it is strictly necessary.

Having determined with precision, in this chapter, the philosophical character of the calculus of indirect functions, according to the principal fundamental conceptions of which it admits, we have next to consider, in the following chapter, the logical division and the general composition of this calculus.

CHAPTER IV.

THE DIFFERENTIAL AND INTEGRAL CALCULUS.

ITS TWO FUNDAMENTAL DIVISIONS.

THE *calculus of indirect functions*, in accordance with the considerations explained in the preceding chapter, is necessarily divided into two parts (or, more properly, is decomposed into two different *calculi* entirely distinct, although intimately connected by their nature), according as it is proposed to find the relations between the auxiliary magnitudes (the introduction of which constitutes the general spirit of this calculus) by means of the relations between the corresponding primitive magnitudes; or, conversely, to try to discover these direct equations by means of the indirect equations originally established. Such is, in fact, constantly the double object of the transcendental analysis.

These two systems have received different names, according to the point of view under which this analysis has been regarded. The infinitesimal method, properly so called, having been the most generally employed for the reasons which have been given, almost all geometers employ habitually the denominations of *Differential Calculus* and of *Integral Calculus*, established by Leibnitz, and which are, in fact, very rational consequences of his conception. Newton, in accordance with his method, named the first the *Calculus of Fluxions*, and the second the *Calculus of Fluents*, expressions which were commonly employed in England. Finally, following the eminently philosophical theory founded by Lagrange, one would be called the *Calculus of Derived Functions,* and the other the *Calculus of Primitive Functions.* I will continue to make use of the terms of Leibnitz, as being more convenient for the formation of secondary expressions, although I ought, in accordance with the suggestions made in the preceding chapter, to employ concurrently all the different conceptions, approaching as nearly as possible to that of Lagrange.

THEIR RELATIONS TO EACH OTHER.

The differential calculus is evidently the logical basis of the integral calculus; for we do not and cannot know how to integrate directly any other differential expressions than those produced by the differentiation of the ten simple functions which constitute the general elements of our analysis. The art of integration consists, then, essentially in bringing all the other cases, as far as is possible, to finally depend on only this small number of fundamental integrations.

In considering the whole body of the transcendental analysis, as I have characterized it in the preceding chapter, it is not at first apparent what can be the peculiar utility of the differential calculus, independently of this necessary relation with the integral calculus, which seems as if it must be, by itself, the only one directly indispensable. In fact, the elimination of the *infinitesimals* or of the *derivatives,* introduced as auxiliaries to facilitate the establishment of equations, constituting, as we have seen, the final and invariable object of the calculus of indirect functions, it is natural to think that the calculus which teaches how to deduce from the equations between these auxiliary magnitudes, those which exist between the primitive magnitudes themselves, ought strictly to suffice for the general wants of the transcendental analysis without our perceiving, at the first glance, what special and constant part the solution of the inverse question can have in such an analysis. It would be a real error, though a common one, to assign to the differential calculus, in order to explain its peculiar, direct, and necessary influence, the destination of forming the differential equations, from which the integral calculus then enables us to arrive at the finite equations; for the primitive formation of differential equations is not and cannot be, properly speaking, the object of any calculus, since, on the contrary, it forms by its nature the indispensable starting point of any calculus whatever. How, in particular, could the differential calculus, which in itself is reduced to teaching the means of *differentiating* the different equations, be a general procedure for establishing them? That which in every application of the transcendental analysis really facilitates the formation of equations, is the infinitesimal *method*, and not the infinitesimal *calculus*, which is perfectly distinct from it, although it is its indispensable complement. Such a consideration would, then, give a false idea of the special destination which characterizes the differential calculus in the general system of the transcendental analysis.

But we should nevertheless very imperfectly conceive the real peculiar importance of this first branch of the calculus of indirect functions, if we saw in it only a simple preliminary labour, having no other general and essential object than to prepare indispensable foundations for the integral calculus. As the ideas on this matter are generally confused, I think that I ought here to explain in a summary manner this important relation as I view it, and to show that in every application of the transcendental analysis a primary, direct, and necessary part is constantly assigned to the differential calculus.

1. *Use of the Differential Calculus as preparatory to that of the Integral.* In forming the differential equations of any phenomenon whatever, it is very seldom that we limit ourselves to introduce differentially only those magnitudes whose relations are sought. To impose that condition would be to uselessly diminish the resources presented by the transcendental analysis for the expression of the mathematical laws of phenomena. Most frequently we introduce into the primitive equations, through their differentials, other magnitudes whose relations are already known or supposed to be so, and without the consideration of which it would be frequently impossible to establish equations. Thus, for example, in the general problem of the rectification of curves, the differential equation,

$$ds^2=dy^2+dx^2, \text{ or } ds^2=dx^2+dy^2+dz^2,$$

is not only established between the desired function s and the independent variable x, to which it is referred, but, at the same time, there have been introduced, as indispensable intermediaries, the differentials of one or two other functions, y and z, which are among the data of the problem; it would not have been possible to form directly the equation between ds and dx, which would, besides, be peculiar to each curve considered. It is the same for most questions. Now in these cases it is evident that the differential equation is not immediately suitable for integration. It is previously necessary that the differentials of the functions supposed to be known, which have been employed as intermediaries, should be entirely eliminated, in order that equations may be obtained between the differentials of the functions which alone are sought and those of the really independent variables, after which the question depends on only the integral calculus. Now this preparatory elimination of certain differentials, in order to reduce the infinitesimals to the smallest number possible, belongs simply to the differential calculus; for it must evidently be done by determining, by means of the equations between the functions supposed to be known, taken as intermediaries, the relations of their differentials, which is merely a question of differentiation. Thus, for example, in the case of rectifications, it will be first necessary to calculate dy, or dy and dz, by differentiating the equation or the equations of each curve proposed; after eliminating these expressions, the general differential formula above enunciated will then contain only ds and dx; having arrived at this point, the elimination of the infinitesimals can be completed only by the integral calculus.

Such is, then, the general office necessarily belonging to the differential calculus in the complete solution of the questions which exact the employment of the transcendental analysis; to produce, as far as is possible, the elimination of the infinitesimals, that is, to reduce in each case the primitive differential equations so that they shall contain only the differentials of the really independent variables, and those of the functions sought, by causing to disappear, by elimination, the differentials of all the other known functions which may have been taken as intermediaries at the time of the formation of the differential equations of the problem which is under consideration.

2. *Employment of the Differential Calculus alone.* For certain questions, which, although few in number, have none the less, as we shall see hereafter, a very great importance, the magnitudes which are sought enter directly, and not by their differentials, into the primitive differential equations, which then contain differentially only the different known functions employed as intermediaries, in accordance with the preceding explanation. These cases are the most favourable of all; for it is evident that the differential calculus is then entirely sufficient for the complete elimination of the infinitesimals, without the question giving rise to any integration. This is what occurs, for example, in the problem of *tangents* in geometry; in that of *velocities* in mechanics, &c.

3. *Employment of the Integral Calculus alone.* Finally, some other questions, the number of which is also very small, but the importance of which is no less great, present a second exceptional case, which is in its nature exactly the converse of the preceding. They are those in which the differential equations are found to be immediately ready for integration, because they contain, at their first formation, only the infinitesimals which relate to the functions sought, or to the really independent variables, without its being necessary to introduce, differentially, other functions as intermediaries. If in these new cases we introduce these last functions, since, by hypothesis, they will enter directly and not by their differentials, ordinary algebra will suffice to eliminate them, and to bring the question to depend on only the integral calculus. The differential calculus will then have no

special part in the complete solution of the problem, which will depend entirely upon the integral calculus. The general question of *quadratures* offers an important example of this, for the differential equation being then $dA=ydx$, will become immediately fit for integration as soon as we shall have eliminated, by means of the equation of the proposed curve, the intermediary function y, which does not enter into it differentially. The same circumstances exist in the problem of *cubatures*, and in some others equally important.

Three classes of Questions hence resulting. As a general result of the previous considerations, it is then necessary to divide into three classes the mathematical questions which require the use of the transcendental analysis; the *first* class comprises the problems susceptible of being entirely resolved by means of the differential calculus alone, without any need of the integral calculus; the *second*, those which are, on the contrary, entirely dependent upon the integral calculus, without the differential calculus having any part in their solution; lastly, in the *third* and the most extensive, which constitutes the normal case, the two others being only exceptional, the differential and the integral calculus have each in their turn a distinct and necessary part in the complete solution of the problem, the former making the primitive differential equations undergo a preparation which is indispensable for the application of the latter. Such are exactly their general relations, of which too indefinite and inexact ideas are generally formed.

Let us now take a general survey of the logical composition of each calculus, beginning with the differential.

THE DIFFERENTIAL CALCULUS.

In the exposition of the transcendental analysis, it is customary to intermingle with the purely analytical part (which reduces itself to the treatment of the abstract principles of differentiation and integration) the study of its different principal applications, especially those which concern geometry. This confusion of ideas, which is a consequence of the actual manner in which the science has been developed, presents, in the dogmatic point of view, serious inconveniences in this respect, that it makes it difficult properly to conceive either analysis or geometry. Having to consider here the most rational co-ordination which is possible, I shall include, in the following sketch, only the calculus of indirect functions properly so called, reserving for the portion of this volume which relates to the philosophical study of *concrete* mathematics the general examination of its great geometrical and mechanical applications.

Two Cases: explicit and implicit Functions. The fundamental division of the differential calculus, or of the general subject of differentiation, consists in distinguishing two cases, according as the analytical functions which are to be differentiated are *explicit* or *implicit;* from which flow two parts ordinarily designated by the names of differentiation *of formulas* and differentiation *of equations.* It is easy to understand, *à priori*, the importance of this classification. In fact, such a distinction would be illusory if the ordinary analysis was perfect; that is, if we knew how to resolve all equations algebraically, for then it would be possible to render every *implicit* function *explicit;* and, by differentiating it in that state alone, the second part of the differential calculus would be immediately comprised in the first, without giving rise to any new difficulty. But the algebraical resolution of equations being, as we have seen, still almost in its infancy, and as yet impossible for most cases, it is plain that the case is very different, since we have, properly speaking, to differentiate a function without knowing it, although it is determinate. The differentiation of implicit functions constitutes then, by its nature, a question truly distinct from that presented by explicit functions, and necessarily more complicated. It is thus evident that we must commence with the differentiation of formulas, and reduce the differentiation of equations to this primary case by certain invariable analytical considerations, which need not be here mentioned.

These two general cases of differentiation are also distinct in another point of view equally necessary, and too important to be left unnoticed. The relation which is obtained between the differentials is constantly more indirect, in comparison with that of the finite quantities, in the differentiation of implicit functions than in that of explicit functions. We know, in fact, from the considerations presented by Lagrange on the general formation of differential equations, that, on the one hand, the same primitive equation may give rise to a greater or less number of derived equations of very different forms, although at bottom equivalent, depending upon which of the arbitrary constants is eliminated, which is not the case in the differentiation of explicit formulas; and that, on the other hand, the unlimited system of the different primitive equations, which correspond to the same derived equation, presents a much more profound analytical variety than that of the different functions, which admit of one same explicit differential, and which are distinguished from each other only by a constant term. Implicit functions must therefore be regarded as being in reality still more modified by differentiation than explicit functions. We shall again meet with this consideration relatively to the integral calculus, where it acquires a preponderant importance.

Two Sub-cases: A single Variable or several Variables. Each of the two fundamental parts of the Differential Calculus is subdivided into two very distinct theories, according as we are required to differentiate functions of a single variable or functions of several independent variables. This second case is, by its nature, quite distinct from the first, and evidently presents more complication, even in considering only explicit functions, and still more those which are implicit. As to the rest, one of these cases is deduced from the other in a general manner, by the aid of an invariable and very simple principle, which consists in regarding the total differential of a function which is produced by the simultaneous increments of the different independent variables which it contains, as the sum of the partial differentials which would be produced by the separate increment of each variable in turn, if all the others were constant. It is necessary, besides, carefully to remark, in connection with this subject, a new idea which is introduced by the distinction of functions into those of one variable and of several; it is the consideration of these different special derived functions, relating to each variable separately, and the number of which increases more and more in proportion as the order of the derivation becomes higher, and also when the variables become more numerous. It results from this that the differential relations belonging to functions of several variables are, by their nature, both much more indirect, and especially much more indeterminate, than those relating to functions of a single variable. This is most apparent in the case of implicit functions, in which, in the place of the simple arbitrary constants which elimination causes to disappear when we form the proper differential equations for functions of a single variable, it is the arbitrary functions of the proposed variables which are then eliminated; whence must result special difficulties when these equations come to be integrated.

Finally, to complete this summary sketch of the different essential parts of the differential calculus proper, I should add, that in the differentiation of implicit functions, whether of a single variable or of several, it is necessary to make another distinction; that of the case in which it is required to differentiate at once different functions of this kind, *combined* in certain primitive equations, from that in which all these functions are *separate*.

The functions are evidently, in fact, still more implicit in the first case than in the second, if we consider that the same imperfection of ordinary analysis, which forbids our converting every implicit function into an equivalent explicit function, in like manner renders us unable to separate the functions which enter simultaneously into any system of equations. It is then necessary to differentiate, not only without knowing how to resolve the primitive equations, but even without being able to effect the proper eliminations among them, thus producing a new difficulty.

Reduction of the whole to the Differentiation of the ten elementary Functions. Such, then, are the natural connection and the logical distribution of the different principal theories which compose the general system of differentiation. Since the differentiation of implicit functions is deduced from that of explicit functions by a single constant principle, and the differentiation of functions of several variables is reduced by another fixed principle to that of functions of a single variable, the whole of the differential calculus is finally found to rest upon the differentiation of explicit functions with a single variable, the only one which is ever executed directly. Now it is easy to understand that this first theory, the necessary basis of the entire system, consists simply in the differentiation of the ten simple functions, which are the uniform elements of all our analytical combinations, and the list of which has been given in the first chapter, on page 51; for the differentiation of compound functions is evidently deduced, in an immediate and necessary manner, from that of the simple functions which compose them. It is, then, to the knowledge of these ten fundamental differentials, and to that of the two general principles just mentioned, which bring under it all the other possible cases, that the whole system of differentiation is properly reduced. We see, by the combination of these different considerations, how simple and how perfect is the entire system of the differential calculus. It certainly constitutes, in its logical relations, the most interesting spectacle which mathematical analysis can present to our understanding.

Transformation of derived Functions for new Variables. The general sketch which I have just summarily drawn would nevertheless present an important deficiency, if I did not here distinctly indicate a final theory, which forms, by its nature, the indispensable complement of the system of differentiation. It is that which has for its object the constant transformation of derived functions, as a result of determinate changes in the independent variables, whence results the possibility of referring to new variables all the general differential formulas primitively established for others. This question is now resolved in the most complete and the most simple manner, as are all those of which the differential calculus is composed. It is easy to conceive the general importance which it must have in any of the applications of the transcendental analysis, the fundamental resources of which it may be considered as augmenting, by permitting us to choose (in order to form the differential equations, in the first place, with more ease) that system of independent variables which may appear to be the most advantageous, although it is not to be finally retained. It is thus, for example, that most of the principal questions of geometry are resolved much more easily by referring the lines and surfaces to *rectilinear* co-ordinates, and that we may, nevertheless, have occasion to express these lines, etc., analytically by the aid of *polar* co-ordinates, or in any other manner. We will then be able to commence the differential solution of the problem by employing the rectilinear system, but only as an intermediate step, from which, by the general theory here referred to, we can pass to the final system, which sometimes could not have been considered directly.

Different Orders of Differentiation. In the logical classification of the differential calculus which has just been given, some may be inclined to suggest a serious omission, since I have not subdivided each of its four essential parts according to another general consideration, which seems at first view very important; namely, that of the higher or lower order of differentiation. But it is easy to understand that this distinction has no real influence in the differential calculus, inasmuch as it does not give rise to any new difficulty. If, indeed, the differential calculus was not rigorously complete, that is, if we did not know how to differentiate at will any function whatever, the differentiation to the second or higher order of each determinate function might engender special difficulties. But the perfect universality of the differential calculus plainly gives us the assurance of being able to differentiate, to any order whatever, all known functions whatever, the question reducing itself to a constantly repeated differentiation of the first order. This distinction, unimportant as it is for the differential calculus, acquires, however, a very great importance in the integral calculus, on account of the extreme imperfection of the latter.

Analytical Applications. Finally, though this is not the place to consider the various applications of the differential calculus, yet an exception may be made for those which consist in the solution of questions which are purely analytical, which ought, indeed, to be logically treated in continuation of a system of differentiation, because of the evident homogeneity of the considerations involved. These questions may be reduced to three essential ones.

Firstly, the *development into series* of functions of one or more variables, or, more generally, the transformation of functions, which constitutes the most beautiful and the most important application of the differential calculus to general analysis, and which comprises, besides the fundamental series discovered by Taylor, the remarkable series discovered by Maclaurin, John Bernouilli, Lagrange, &c.:

Secondly, the general *theory of maxima and minima* values for any functions whatever, of one or more variables; one of the most interesting problems which analysis can present, however elementary it may now have become, and to the complete solution of which the differential calculus naturally applies:

Thirdly, the general determination of the true value of functions which present themselves under an *indeterminate* appearance for certain hypotheses made on the values of the corresponding variables; which is the least extensive and the least important of the three.

The first question is certainly the principal one in all points of view; it is also the most susceptible of receiving a new extension hereafter, especially by conceiving, in a broader manner than has yet been done, the employment of the differential calculus in the transformation of functions, on which subject Lagrange has left some valuable hints.

Having thus summarily, though perhaps too briefly, considered the chief points in the differential calculus, I now proceed to an equally rapid exposition of a systematic outline of the Integral Calculus, properly so called, that is, the abstract subject of integration.

THE INTEGRAL CALCULUS.

Its Fundamental Division. The fundamental division of the Integral Calculus is founded on the same principle as that of the Differential Calculus, in distinguishing the integration of *explicit* differential formulas, and the integration of *implicit* differentials or of differential equations. The separation of these two cases is even much more profound in relation to integration than to differentiation. In the differential calculus, in fact, this distinction rests, as we have seen, only on the extreme imperfection of ordinary analysis. But, on the other hand, it is easy to see that, even though all equations could be algebraically resolved, differential equations would none the less constitute a case of integration quite distinct from that presented by the explicit differential formulas; for, limiting ourselves, for the sake of simplicity, to the first order, and to a single function y of a single variable x, if we suppose any differential equation between x, y, and $\frac{dy}{dx}$, to be resolved with reference to $\frac{dy}{dx}$, the expression of the derived function being then generally found to contain the primitive function itself, which is the object of the inquiry, the question of integration will not have at all changed its nature, and the solution will not really have made any other progress than that of having brought the proposed differential equation to be of only the first degree relatively to the derived function, which is in itself of little importance. The differential would not then be determined in a manner much less *implicit* than before, as regards the integration, which would continue to present essentially the same characteristic difficulty. The algebraic resolution of equations could not make the case which we are considering come within the simple integration of explicit differentials, except in the special cases in which the proposed differential equation did not contain the primitive function itself, which would consequently permit us, by resolving it, to find $\frac{dy}{dx}$ in terms of x only, and thus to reduce the question to the class of quadratures. Still greater difficulties would evidently be found in differential equations of higher orders, or containing simultaneously different functions of several independent variables.

The integration of differential equations is then necessarily more complicated than that of explicit differentials, by the elaboration of which last the integral calculus has been created, and upon which the others have been made to depend as far as it has been possible. All the various analytical methods which have been proposed for integrating differential equations, whether it be the separation of the variables, the method of multipliers, &c., have in fact for their object to reduce these integrations to those of differential formulas, the only one which, by its nature, can be undertaken directly. Unfortunately, imperfect as is still this necessary base of the whole integral calculus, the art of reducing to it the integration of differential equations is still less advanced.

Subdivisions: one variable or several. Each of these two fundamental branches of the integral calculus is next subdivided into two others (as in the differential calculus, and for precisely analogous reasons), according as we consider functions with a *single variable*, or functions with *several independent variables*.

This distinction is, like the preceding one, still more important for integration than for differentiation. This is especially remarkable in reference to differential equations. Indeed, those which depend on several independent variables may evidently present this characteristic and much more serious difficulty, that the desired function may be differentially defined by a simple relation between its different special derivatives relative to the different variables taken separately. Hence results the most difficult and also the most extensive branch of the integral calculus, which is commonly named the *Integral Calculus of partial differences*, created by D'Alembert, and in which, according to the just appreciation of Lagrange, geometers ought to have seen a really new calculus, the philosophical character of which has not yet been determined with sufficient exactness. A very striking difference between this case and that of equations with a single independent variable consists, as has been already observed, in the arbitrary functions which take the place of the simple arbitrary constants, in order to give to the corresponding integrals all the proper generality.

It is scarcely necessary to say that this higher branch of transcendental analysis is still entirely in its infancy, since, even in the most simple case, that of an equation of the first order between the partial derivatives of a single function with two independent variables, we are not yet completely able to reduce the integration to that of the ordinary differential equations. The integration of functions of several variables is much farther advanced in the case (infinitely more simple indeed) in which it has to do with only explicit differential formulas. We can then, in fact, when these formulas fulfil the necessary conditions of integrability, always reduce their integration to quadratures.

Other Subdivisions: different Orders of Differentiation. A new general distinction, applicable as a subdivision to the integration of explicit or implicit differentials, with one variable or several, is drawn from the *higher or lower order of the differentials:* a distinction which, as we have above remarked, does not give rise to any special question in the differential calculus.

Relatively to *explicit differentials*, whether of one variable or of several, the necessity of distinguishing their different orders belongs only to the extreme imperfection of the integral calculus. In fact, if we could always integrate every differential formula of the first order, the integration of a formula of the second order, or of any other, would evidently not form a new question, since, by integrating it at first in the first degree, we would arrive at the differential expression of the immediately preceding order, from which, by a suitable series of analogous integrations, we would be certain of finally arriving at the primitive function, the final object of these operations. But the little knowledge which we possess on integration of even the first order causes quite another state of affairs, so that a higher order of differentials produces new difficulties; for, having differential formulas of any order above the first, it may happen that we may be able to integrate them, either once, or several times in succession, and that we may still be unable to go back to the primitive functions, if these preliminary labours have produced, for the differentials of a lower order, expressions whose integrals are not known. This circumstance must occur so much the oftener (the number of known integrals being still very small), seeing that these successive integrals are generally very different functions from the derivatives which have produced them.

With reference to *implicit differentials*, the distinction of orders is still more important; for, besides the preceding reason, the influence of which is evidently analogous in this case, and is even greater, it is easy to perceive that the higher order of the differential equations necessarily gives rise to questions of a new nature. In fact, even if we could integrate every equation of the first order relating to a single function, that would not be sufficient for obtaining the final integral of an equation of any order whatever, inasmuch as every differential equation is not reducible to that of an immediately inferior order. Thus, for example, if we have given any relation between x, y, $\frac{dx}{dy}$, and $\frac{d^2y}{dx^2}$, to determine a function y of a variable x, we shall not be able to deduce from it at once, after effecting a first integration, the corresponding differential relation between x, y, and $\frac{dy}{dx}$, from which, by a second integration, we could ascend to the primitive equations. This would not necessarily take place, at least without introducing new auxiliary functions, unless the proposed equation of the second order did not contain the required function y, together with its derivatives. As a general principle, differential equations will have to be regarded as presenting cases which are more and more *implicit*, as they are of a higher order, and which cannot be made to depend on one another except by special methods, the investigation of which consequently forms a new class of questions, with respect to which we as yet know scarcely any thing, even for functions of a single variable.*

Another equivalent distinction. Still farther, when we examine more profoundly this distinction of different orders of differential equations, we find that it can be always made to come under a final general distinction, relative to differential equations, which remains to be noticed. Differential equations with one or more independent variables may contain simply a single function, or (in a case evidently more complicated and more implicit, which corresponds to the differentiation of simultaneous implicit functions) we may have to determine at the same time several functions from the differential equations in which they are found united, together with their different derivatives. It is clear that such a state of the question necessarily presents a new special difficulty, that of separating the different functions desired, by forming for each, from the proposed differential equations, an isolated differential equation which does not contain the other functions or their derivatives. This preliminary labour, which is analogous to the elimination of algebra, is evidently indispensable before attempting any direct integration, since we cannot undertake generally (except by special artifices which are very rarely applicable) to determine directly several distinct functions at once.

Now it is easy to establish the exact and necessary coincidence of this new distinction with the preceding

---

* The only important case of this class which has thus far been completely treated is the general integration of *linear* equations of any order whatever, with constant coefficients. Even this case finally depends on the algebraic resolution of equations of a degree equal to the order of differentiation.

one respecting the order of differential equations. We know, in fact, that the general method for isolating functions in simultaneous differential equations consists essentially in forming differential equations, separately in relation to each function, and of an order equal to the sum of all those of the different proposed equations. This transformation can always be effected. On the other hand, every differential equation of any order in relation to a single function might evidently always be reduced to the first order, by introducing a suitable number of auxiliary differential equations, containing at the same time the different anterior derivatives regarded as new functions to be determined. This method has, indeed, sometimes been actually employed with success, though it is not the natural one.

Here, then, are two necessarily equivalent orders of conditions in the general theory of differential equations; the simultaneousness of a greater or smaller number of functions, and the higher or lower order of differentiation of a single function. By augmenting the order of the differential equations, we can isolate all the functions; and, by artificially multiplying the number of the functions, we can reduce all the equations to the first order. There is, consequently, in both cases, only one and the same difficulty from two different points of sight. But, however we may conceive it, this new difficulty is none the less real, and constitutes none the less, by its nature, a marked separation between the integration of equations of the first order and that of equations of a higher order. I prefer to indicate the distinction under this last form as being more simple, more general, and more logical.

Quadratures. From the different considerations which have been indicated respecting the logical dependence of the various principal parts of the integral calculus, we see that the integration of explicit differential formulas of the first order and of a single variable is the necessary basis of all other integrations, which we never succeed in effecting but so far as we reduce them to this elementary case, evidently the only one which, by its nature, is capable of being treated directly. This simple fundamental integration is often designated by the convenient expression of *quadratures*, seeing that every integral of this kind, $Sf(x)dx$, may, in fact, be regarded as representing the area of a curve, the equation of which in rectilinear co-ordinates would be $y=f(x)$. Such a class of questions corresponds, in the differential calculus, to the elementary case of the differentiation of explicit functions of a single variable. But the integral question is, by its nature, very differently complicated, and especially much more extensive than the differential question. This latter is, in fact, necessarily reduced, as we have seen, to the differentiation of the ten simple functions, the elements of all which are considered in analysis. On the other hand, the integration of compound functions does not necessarily follow from that of the simple functions, each combination of which may present special difficulties with respect to the integral calculus. Hence results the naturally indefinite extent, and the so varied complication of the question of *quadratures*, upon which, in spite of all the efforts of analysts, we still possess so little complete knowledge.

In decomposing this question, as is natural, according to the different forms which may be assumed by the derivative function, we distinguish the case of *algebraic* functions and that of *transcendental* functions.

Integration of Transcendental Functions. The truly analytical integration of transcendental functions is as yet very little advanced, whether for *exponential*, or for *logarithmic*, or for *circular* functions. But a very small number of cases of these three different kinds have as yet been treated, and those chosen from among the simplest; and still the necessary calculations are in most cases extremely laborious. A circumstance which we ought particularly to remark in its philosophical connection is, that the different procedures of quadrature have no relation to any general view of integration, and consist of simple artifices very incoherent with each other, and very numerous, because of the very limited extent of each.

One of these artifices should, however, here be noticed, which, without being really a method of integration, is nevertheless remarkable for its generality; it is the procedure invented by John Bernouilli, and known under the name of *integration by parts*, by means of which every integral may be reduced to another which is sometimes found to be more easy to be obtained. This ingenious relation deserves to be noticed for another reason, as having suggested the first idea of that transformation of integrals yet unknown, which has lately received a greater extension, and of which M. Fourier especially has made so new and important a use in the analytical questions produced by the theory of heat.

Integration of Algebraic Functions. As to the integration of algebraic functions, it is farther advanced. However, we know scarcely any thing in relation to irrational functions, the integrals of which have been obtained only in extremely limited cases, and particularly by rendering them rational. The integration of rational functions is thus far the only theory of the integral calculus which has admitted of being treated in a truly complete manner; in a logical point of view, it forms, then, its most satisfactory part, but perhaps also the least important. It is even essential to remark, in order to have a just idea of the extreme imperfection of the integral calculus, that this case, limited as it is, is not entirely resolved except for what properly concerns integration viewed in an abstract manner; for, in the execution, the theory finds its progress most frequently quite stopped, independently of the complication of the calculations, by the imperfection of ordinary analysis, seeing that it makes the integration finally depend upon the algebraic resolution of equations, which greatly limits its use.

To grasp in a general manner the spirit of the different procedures which are employed in quadratures, we must observe that, by their nature, they can be primitively founded only on the differentiation of the ten simple functions. The results of this, conversely considered, establish as many direct theorems of the integral calculus, the only ones which can be directly known. All the art of integration afterwards consists, as has been said in the beginning of this chapter, in reducing all the other quadratures, so far as is possible, to this small number of elementary ones, which unhappily we are in most cases unable to effect.

Singular Solutions. In this systematic enumeration of the various essential parts of the integral calculus, considered in their logical relations, I have designedly neglected (in order not to break the chain of sequence) to consider a very important theory, which forms implicitly a portion of the general theory of the integration of differential equations, but which I ought here to notice separately, as being, so to speak, outside of the integral calculus, and being nevertheless of the greatest interest, both by its logical perfection and by the extent of its applications. I refer to what are called *Singular Solutions* of differential equations, called sometimes, but improperly, *particular* solutions, which have been the subject of very remarkable investigations by Euler and Laplace, and of which Lagrange especially has presented such a beautiful and simple general theory. Clairaut, who first had occasion to remark their existence, saw in them a paradox of the integral calculus, since these solutions have the peculiarity of satisfying the differential equations without being comprised in the corresponding general integrals. Lagrange has since explained this paradox in the most ingenious and most satisfactory manner, by showing how such solutions are always derived from the general integral by the variation of the arbitrary constants. He was also the first to suitably appreciate the importance of this theory, and it is with good reason that he devoted to it so full a development in his "Calculus of Functions." In a logical point of view, this theory deserves all our attention by the character of perfect generality which it admits of, since Lagrange has given invariable and very simple procedures for finding the *singular* solution of any differential equation which is susceptible of it; and, what is no less remarkable, these procedures require no integration, consisting only of differentiations, and are therefore always applicable. Differentiation has thus become, by a happy artifice, a means of compensating, in certain circumstances, for the imperfection of the integral calculus. Indeed, certain problems especially require, by their nature, the knowledge of these *singular* solutions; such, for example, in geometry, are all the questions in which a curve is to be determined from any property of its tangent or its osculating circle. In all cases of this kind, after having expressed this property by a differential equation, it will be, in its analytical relations, the *singular* equation which will form the most important object of the inquiry, since it alone will represent the required curve; the general integral, which thenceforth it becomes unnecessary to know, designating only the system of the tangents, or of the osculating circles of this curve. We may hence easily understand all the importance of this theory, which seems to me to be not as yet sufficiently appreciated by most geometers.

Definite Integrals. Finally, to complete our review of the vast collection of analytical researches of which is composed the integral calculus, properly so called, there remains to be mentioned one theory, very important in all the applications of the transcendental analysis, which I have had to leave outside of the system, as not being really destined for veritable integration, and proposing, on the contrary, to supply the place of the knowledge of truly analytical integrals, which are most generally unknown. I refer to the determination of *definite integrals.*

The expression, always possible, of integrals in infinite series, may at first be viewed as a happy general means of compensating for the extreme imperfection of the integral calculus. But the employment of such series, because of their complication, and of the difficulty of discovering the law of their terms, is commonly of only moderate utility in the algebraic point of view, although sometimes very essential relations have been thence deduced. It is particularly in the arithmetical point of view that this procedure acquires a great importance, as a means of calculating what are called *definite integrals*, that is, the values of the required functions for certain determinate values of the corresponding variables.

An inquiry of this nature exactly corresponds, in transcendental analysis, to the numerical resolution of equations in ordinary analysis. Being generally unable to obtain the veritable integral—named by opposition the *general* or *indefinite* integral; that is, the function which, differentiated, has produced the proposed differential formula—analysts have been obliged to employ themselves in determining at least, without knowing this function, the particular numerical values which it would take on assigning certain designated values to the variables. This is evidently resolving the arithmetical question without having previously resolved the corresponding algebraic one, which most generally is the most important one. Such an analysis is, then, by its nature, as imperfect as we have seen the numerical resolution of equations to be. It presents, like this last, a vicious confusion of arithmetical and algebraic considerations, whence result analogous inconveniences both in the purely logical point of view and in the applications. We need not here repeat the considerations suggested in our third chapter. But it will be understood that, unable as we almost always are to obtain the true integrals, it is of the highest importance to have been able to obtain this solution, incomplete and necessarily insufficient as it is. Now this has been fortunately attained at the present day for all cases, the determination of the value of definite integrals having been reduced to entirely general methods, which leave nothing to desire, in a great number of cases, but less complication in the calculations, an object towards which are at present directed all the special transformations of analysts. Regarding now this sort of *transcendental arithmetic* as perfect, the difficulty in the applications is essentially reduced to making the proposed research depend, finally, on a simple determination of definite integrals, which evidently cannot always be possible, whatever analytical skill may be employed in effecting such a transformation.

Prospects of the Integral Calculus. From the considerations indicated in this chapter, we see that, while the differential calculus constitutes by its nature a limited and perfect system, to which nothing essential remains to be added, the integral calculus, or the simple system of integration, presents necessarily an inexhaustible field for the activity of the human mind, independently of the indefinite applications of which the transcendental analysis is evidently susceptible. The general argument by which I have endeavoured, in the second chapter, to make apparent the impossibility of ever discovering the algebraic solution of equations of any degree and form whatsoever, has undoubtedly infinitely more force with regard to the search for a single method of integration, invariably applicable to all cases. "It is," says Lagrange, "one of those problems whose general solution we cannot hope for." The more we meditate on this subject, the more we will be convinced that such a research is utterly chimerical, as being far above the feeble reach of our intelligence; although the labours of geometers must certainly augment hereafter the amount of our knowledge respecting integration, and thus create methods of greater generality. The transcendental analysis is still too near its origin—there is especially too little time since it has been conceived in a truly rational manner—for us now to be able to have a correct idea of what it will hereafter become. But, whatever should be our legitimate hopes, let us not forget to consider, before all, the limits which are imposed by our intellectual constitution, and which, though not susceptible of a precise determination, have none the less an incontestable reality.

I am induced to think that, when geometers shall have exhausted the most important applications of our present transcendental analysis, instead of striving to impress upon it, as now conceived, a chimerical perfection, they will rather create new resources by changing the mode of derivation of the auxiliary quantities introduced in order to facilitate the establishment of equations, and the formation of which might follow an infinity of other laws besides the very simple relation which has been chosen, according to the conception suggested in the first chapter. The resources of this nature appear to me susceptible of a much greater fecundity than those which would consist of merely pushing farther our present calculus of indirect functions. It is a suggestion which I submit to the geometers who have turned their thoughts towards the general philosophy of analysis.

Finally, although, in the summary exposition which was the object of this chapter, I have had to exhibit the condition of extreme imperfection which still belongs to the integral calculus, the student would have a false idea of the general resources of the transcendental analysis if he gave that consideration too great an importance. It is with it, indeed, as with ordinary analysis, in which a very small amount of fundamental knowledge respecting the resolution of equations has been employed with an immense degree of utility. Little advanced as geometers really are as yet in the science of integrations, they have nevertheless obtained, from their scanty abstract conceptions, the solution of a multitude of questions of the first importance in geometry, in mechanics, in thermology, &c. The philosophical explanation of this double general fact results from the necessarily preponderating importance and grasp of *abstract* branches of knowledge, the least of which is naturally found to correspond to a crowd of *concrete* researches, man having no other resource for the successive extension of his intellectual means than in the consideration of ideas more and more abstract, and still positive.

In order to finish the complete exposition of the philosophical character of the transcendental analysis, there remains to be considered a final conception, by which the immortal Lagrange has rendered this analysis still better adapted to facilitate the establishment of equations in the most difficult problems, by considering a class of equations still more *indirect* than the ordinary differential equations. It is the *Calculus*, or, rather, the *Method of Variations;* the general appreciation of which will be our next subject.

CHAPTER V.

THE CALCULUS OF VARIATIONS.

In order to grasp with more ease the philosophical character of the *Method of Variations*, it will be well to begin by considering in a summary manner the special nature of the problems, the general resolution of which has rendered necessary the formation of this hyper-transcendental analysis. It is still too near its origin, and its applications have been too few, to allow us to obtain a sufficiently clear general idea of it from a purely abstract exposition of its fundamental theory.

PROBLEMS GIVING RISE TO IT.

The mathematical questions which have given birth to the *Calculus of Variations* consist generally in the investigation of the *maxima* and *minima* of certain indeterminate integral formulas, which express the analytical law of such or such a phenomenon of geometry or mechanics, considered independently of any particular subject. Geometers for a long time designated all the questions of this character by the common name of *Isoperimetrical Problems*, which, however, is really suitable to only the smallest number of them.

Ordinary Questions of Màxima and Minima. In the common theory of *maxima* and *minima*, it is proposed to discover, with reference to a given function of one or more variables, what particular values must be assigned to these variables, in order that the corresponding value of the proposed function may be a *maximum* or a *minimum* with respect to those values which immediately precede and follow it; that is, properly speaking, we seek to know at what instant the function ceases to increase and commences to decrease, or reciprocally. The differential calculus is perfectly sufficient, as we know, for the general resolution of this class of questions, by showing that the values of the different variables, which suit either the maximum or minimum, must always reduce to zero the different first derivatives of the given function, taken separately with reference to each independent variable, and by indicating, moreover, a suitable characteristic for distinguishing the maximum from the minimum; consisting, in the case of a function of a single variable, for example, in the derived function of the second order taking a negative value for the maximum, and a positive value for the minimum. Such are the well-known fundamental conditions belonging to the greatest number of cases.

A new Class of Questions. The construction of this general theory having necessarily destroyed the chief interest which questions of this kind had for geometers, they almost immediately rose to the consideration of a new order of problems, at once much more important and of much greater difficulty—those of *isoperimeters.* It is, then, no longer *the values of the variables* belonging to the maximum or the minimum of a given function that it is required to determine. It is *the form of the function itself* which is required to be discovered, from the condition of the maximum or of the minimum of a certain definite integral, merely indicated, which depends upon that function.

Solid of least Resistance. The oldest question of this nature is that of *the solid of least resistance*, treated by Newton in the second book of the Principia, in which he determines what ought to be the meridian curve of a solid of revolution, in order that the resistance experienced by that body in the direction of its axis may be the least possible. But the course pursued by Newton, from the nature of his special method of transcendental analysis, had not a character sufficiently simple, sufficiently general, and especially sufficiently analytical, to attract geometers to this new order of problems. To effect this, the application of the infinitesimal method was needed; and this was done, in 1695, by John Bernouilli, in proposing the celebrated problem of the *Brachystochrone*.

This problem, which afterwards suggested such a long series of analogous questions, consists in determining the curve which a heavy body must follow in order to descend from one point to another in the shortest possible time. Limiting the conditions to the simple fall in a vacuum, the only case which was at first considered, it is easily found that the required curve must be a reversed cycloid with a horizontal base, and with its origin at the highest point. But the question may become singularly complicated, either by taking into account the resistance of the medium, or the change in the intensity of gravity.

Isoperimeters. Although this new class of problems was in the first place furnished by mechanics, it is in geometry that the principal investigations of this character were subsequently made. Thus it was proposed to discover which, among all the curves of the same contour traced between two given points, is that whose area is a maximum or minimum, whence has come the name of *Problem of Isoperimeters;* or it was required that the maximum or minimum should belong to the surface produced by the revolution of the required curve about an axis, or to the corresponding volume; in other cases, it was the vertical height of the center of gravity of the unknown curve, or of the surface and of the volume which it might generate, which was to become a maximum or minimum, &c. Finally, these problems were varied and complicated almost to infinity by the Bernouillis, by Taylor, and especially by Euler, before Lagrange reduced their solution to an abstract and entirely general method, the discovery of which has put a stop to the enthusiasm of geometers for such an order of inquiries. This is not the place for tracing the history of this subject. I have only enumerated some of the simplest principal questions, in order to render apparent the original general object of the method of variations.

Analytical Nature of these Problems. We see that all these problems, considered in an analytical point of view, consist, by their nature, in determining what form a certain unknown function of one or more variables ought to have, in order that such or such an integral, dependent upon that function, shall have, within assigned limits, a value which is a maximum or a minimum with respect to all those which it would take if the required function had any other form whatever.

Thus, for example, in the problem of the *brachystochrone*, it is well known that if $y=f(z)$, $x=\phi(z)$, are the rectilinear equations of the required curve, supposing the axes of x and of y to be horizontal, and the axis of

z to be vertical, the time of the fall of a heavy body in that curve from the point whose ordinate is z_1, to that whose ordinate is z_2, is expressed in general terms by the definite integral

$$\int_{z_2}^{z^1} \sqrt{\frac{1+(f'(z))^2+(\phi'(z))^2}{2gz}}dz.$$

It is, then, necessary to find what the two unknown functions f and ϕ must be, in order that this integral may be a minimum.

In the same way, to demand what is the curve among all plane isoperimetrical curves, which includes the greatest area, is the same thing as to propose to find, among all the functions $f(x)$ which can give a certain constant value to the integral

$$\int dx \sqrt{1+(f'(x))^2},$$

that one which renders the integral $\int f(x)dx$, taken between the same limits, a maximum. It is evidently always so in other questions of this class.

Methods of the older Geometers. In the solutions which geometers before Lagrange gave of these problems, they proposed, in substance, to reduce them to the ordinary theory of maxima and minima. But the means employed to effect this transformation consisted in special simple artifices peculiar to each case, and the discovery of which did not admit of invariable and certain rules, so that every really new question constantly reproduced analogous difficulties, without the solutions previously obtained being really of any essential aid, otherwise than by their discipline and training of the mind. In a word, this branch of mathematics presented, then, the necessary imperfection which always exists when the part common to all questions of the same class has not yet been distinctly grasped in order to be treated in an abstract and thenceforth general manner.

METHOD OF LAGRANGE.

Lagrange, in endeavouring to bring all the different problems of isoperimeters to depend upon a common analysis, organized into a distinct calculus, was led to conceive a new kind of differentiation, to which he has applied the characteristic δ, reserving the characteristic d for the common differentials. These differentials of a new species, which he has designated under the name of *Variations*, consist of the infinitely small increments which the integrals receive, not by virtue of analogous increments on the part of the corresponding variables, as in the ordinary transcendental analysis, but by supposing that the *form* of the function placed under the sign of integration undergoes an infinitely small change. This distinction is easily conceived with reference to curves, in which we see the ordinate, or any other variable of the curve, admit of two sorts of differentials, evidently very different, according as we pass from one point to another infinitely near it on the same curve, or to the corresponding point of the infinitely near curve produced by a certain determinate modification of the first curve.* It is moreover clear, that the relative *variations* of different magnitudes connected with each other by any laws whatever are calculated, all but the characteristic, almost exactly in the same manner as the differentials. Finally,

* Leibnitz had already considered the comparison of one curve with another infinitely near to it, calling it "*Differentiatio de curva in curvam.*" But this comparison had no analogy with the conception of Lagrange, the curves of Leibnitz being embraced in the same general equation, from which they were deduced by the simple change of an arbitrary constant.

from the general notion of *variations* are in like manner deduced the fundamental principles of the algorithm proper to this method, consisting simply in the evidently permissible liberty of transposing at will the characteristics specially appropriated to variations, before or after those which correspond to the ordinary differentials.

This abstract conception having been once formed, Lagrange was able to reduce with ease, and in the most general manner, all the problems of *Isoperimeters* to the simple ordinary theory of *maxima* and *minima*. To obtain a clear idea of this great and happy transformation, we must previously consider an essential distinction which arises in the different questions of isoperimeters.

Two Classes of Questions. These investigations must, in fact, be divided into two general classes, according as the maxima and minima demanded are *absolute* or *relative*, to employ the abridged expressions of geometers.

Questions of the first Class. The *first case* is that in which the indeterminate definite integrals, the maximum or minimum of which is sought, are not subjected, by the nature of the problem, to any condition; as happens, for example, in the problem of the *brachystochrone*, in which the choice is to be made between all imaginable curves. The *second* case takes place when, on the contrary, the variable integrals can vary only according to certain conditions, which usually consist in other definite integrals (which depend, in like manner, upon the required functions) always retaining the same given value; as, for example, in all the geometrical questions relating to real *isoperimetrical* figures, and in which, by the nature of the problem, the integral relating to the length of the curve, or to the area of the surface, must remain constant during the variation of that integral which is the object of the proposed investigation.

The *Calculus of Variations* gives immediately the general solution of questions of the former class; for it evidently follows, from the ordinary theory of maxima and minima, that the required relation must reduce to zero the *variation* of the proposed integral with reference to each independent variable; which gives the condition common to both the maximum and the minimum: and, as a characteristic for distinguishing the one from the other, that the variation of the second order of the same integral must be negative for the maximum and positive for the minimum. Thus, for example, in the problem of the brachystochrone, we will have, in order to determine the nature of the curve sought, the equation of condition

$$\delta\int_{z_2}^{z^1}\sqrt{\frac{1+(f'(z))^2+(\phi'(z))^2}{2gz}}\,dz=0,$$

which, being decomposed into two, with respect to the two unknown functions f and ϕ, which are independent of each other, will completely express the analytical definition of the required curve. The only difficulty peculiar to this new analysis consists in the elimination of the characteristic δ, for which the calculus of variations furnishes invariable and complete rules, founded, in general, on the method of "integration by parts," from which Lagrange has thus derived immense advantage. The constant object of this first analytical elaboration (which this is not the place for treating in detail) is to arrive at real differential equations, which can always be done; and thereby the question comes under the ordinary transcendental analysis, which furnishes the solution, at least so far as to reduce it to pure algebra if the integration can be effected. The general object of the method of variations is to effect this transformation, for which Lagrange has established rules, which are simple, invariable, and certain of success.

Equations of Limits. Among the greatest special advantages of the method of variations, compared with the previous isolated solutions of isoperimetrical problems, is the important consideration of what Lagrange calls *Equations of Limits,* which were entirely neglected before him, though without them the greater part of the particular solutions remained necessarily incomplete. When the limits of the proposed integrals are to be fixed, their variations being zero, there is no occasion for noticing them. But it is no longer so when these limits, instead of being rigorously invariable, are only subjected to certain conditions; as, for example, if the two points between which the required curve is to be traced are not fixed, and have only to remain upon given lines or surfaces. Then it is necessary to pay attention to the variation of their co-ordinates, and to establish between them the relations which correspond to the equations of these lines or of these surfaces.

A more general consideration. This essential consideration is only the final complement of a more general and more important consideration relative to the variations of different independent variables. If these variables are really independent of one another, as when we compare together all the imaginable curves susceptible of being traced between two points, it will be the same with their variations, and, consequently, the terms relating to each of these variations will have to be separately equal to zero in the general equation which expresses the maximum or the minimum. But if, on the contrary, we suppose the variables to be subjected to any fixed conditions, it will be necessary to take notice of the resulting relation between their variations, so that the number of the equations into which this general equation is then decomposed is always equal to only the number of the variables which remain truly independent. It is thus, for example, that instead of seeking for the shortest path between any two points, in choosing it from among all possible ones, it may be proposed to find only what is the shortest among all those which may be taken on any given surface; a question the general solution of which forms certainly one of the most beautiful applications of the method of variations.

Questions of the second Class. Problems in which such modifying conditions are considered approach very nearly, in their nature, to the second general class of applications of the method of variations, characterized above as consisting in the investigation of *relative* maxima and minima. There is, however, this essential difference between the two cases, that in this last the modification is expressed by an integral which depends upon the function sought, while in the other it is designated by a finite equation which is immediately given. It is hence apparent that the investigation of *relative* maxima and minima is constantly and necessarily more complicated than that of *absolute* maxima and minima. Luckily, a very important general theory, discovered by the genius of the great Euler before the invention of the Calculus of Variations, gives a uniform and very simple means of making one of these two classes of questions dependent on the other. It consists in this, that if we add to the integral which is to be a maximum or a minimum, a constant and indeterminate multiple of that one which, by the nature of the problem, is to remain constant, it will be sufficient to seek, by the general method of Lagrange above indicated, the *absolute* maximum or minimum of this whole expression. It can be easily conceived, indeed, that the part of the complete variation which would proceed from the last integral must be equal to zero (because of the constant character of this last) as well as the portion due to the first integral, which disappears by virtue of the maximum or minimum state. These two conditions evidently unite to produce, in that respect, effects exactly alike.

Such is a sketch of the general manner in which the method of variation is applied to all the different questions which compose what is called the *Theory of Isoperimeters.* It will undoubtedly have been remarked in this summary exposition how much use has been made in this new analysis of the second fundamental property of the transcendental analysis noticed in the third chapter, namely, the generality of the infinitesimal expressions for the representation of the same geometrical or mechanical phenomenon, in whatever body it may be considered. Upon this generality, indeed, are founded, by their nature, all the solutions due to the method of variations. If a single formula could not express the length or the area of any curve whatever; if another fixed formula could not designate the time of the fall of a heavy body, according to whatever line it may descend, &c., how would it have been possible to resolve questions which unavoidably require, by their nature, the simultaneous consideration of all the cases which can be determined in each phenomenon by the different subjects which exhibit it.

Other Applications of this Method. Notwithstanding the extreme importance of the theory of isoperimeters, and though the method of variations had at first no other object than the logical and general solution of this order of problems, we should still have but an incomplete idea of this beautiful analysis if we limited its destination to this. In fact, the abstract conception of two distinct natures of differentiation is evidently applicable not only to the cases for which it was created, but also to all those which present, for any reason whatever, two different manners of making the same magnitudes vary. It is in this way that Lagrange himself has made, in his "*Mecanique Analytique*," an extensive and important application of his calculus of variations, by employing it to distinguish the two sorts of changes which are naturally presented by the questions of rational mechanics for the different points which are considered, according as we compare the successive positions which are occupied, in virtue of its motion, by the same point of each body in two consecutive instants, or as we pass from one point of the body to another in the same instant. One of these comparisons produces ordinary differentials; the other gives rise to *variations*, which, there as every where, are only differentials taken under a new point of view. Such is the general acceptation in which we should conceive the Calculus of Variations, in order suitably to appreciate the importance of this admirable logical instrument, the most powerful that the human mind has as yet constructed.

The method of variations being only an immense extension of the general transcendental analysis, I have no need of proving specially that it is susceptible of being considered under the different fundamental points of view which the calculus of indirect functions, considered as a whole, admits of. Lagrange invented the Calculus of Variations in accordance with the infinitesimal conception, and, indeed, long before he undertook the general reconstruction of the transcendental analysis. When he had executed this important reformation, he easily showed how it could also be applied to the Calculus of Variations, which he expounded with all the proper development, according to his theory of derivative functions. But the more that the use of the method of variations is difficult of comprehension, because of the higher degree of abstraction of the ideas considered, the more necessary is it, in its application, to economize the exertions of the mind, by adopting the most direct and rapid analytical conception, namely, that of Leibnitz. Accordingly, Lagrange himself has constantly preferred it in the important use which he has made of the Calculus of Variations in his "Analytical Mechanics." In fact, there does not exist the least hesitation in this respect among geometers.

ITS RELATIONS TO THE ORDINARY CALCULUS.

In order to make as clear as possible the philosophical character of the Calculus of Variations, I think that I should, in conclusion, briefly indicate a consideration which seems to me important. and by which I can approach it to the ordinary transcendental analysis in a higher degree than Lagrange seems to me to have done.*

We noticed in the preceding chapter the formation of the *calculus of partial differences*, created by D'Alembert, as having introduced into the transcendental analysis a new elementary idea; the notion of two kinds of increments, distinct and independent of one another, which a function of two variables may receive by virtue of the change of each variable separately. It is thus that the vertical ordinate of a surface, or any other magnitude which is referred to it, varies in two manners which are quite distinct, and which may follow the most different laws, according as we increase either the one or the other of the two horizontal co-ordinates. Now such a consideration seems to me very nearly allied, by its nature, to that which serves as the general basis of the method of variations. This last, indeed, has in reality done nothing but transfer to the independent variables themselves the peculiar conception which had been already adopted for the functions of these variables; a modification which has remarkably enlarged its use. I think, therefore, that so far as regards merely the fundamental conceptions, we may consider the calculus created by D'Alembert as having established a natural and necessary transition between the ordinary infinitesimal calculus and the calculus of variations; such a derivation of which seems to be adapted to make the general notion more clear and simple.

---

* I propose hereafter to develop this new consideration, in a special work upon the *Calculus of Variations*, intended to present this hyper-transcendental analysis in a new point of view, which I think adapted to extend its general range.

According to the different considerations indicated in this chapter, the method of variations presents itself as the highest degree of perfection which the analysis of indirect functions has yet attained. In its primitive state, this last analysis presented itself as a powerful general means of facilitating the mathematical study of natural phenomena, by introducing, for the expression of their laws, the consideration of auxiliary magnitudes, chosen in such a manner that their relations are necessarily more simple and more easy to obtain than those of the direct magnitudes. But the formation of these differential equations was not supposed to admit of any general and abstract rules. Now the Analysis of Variations, considered in the most philosophical point of view, may be regarded as essentially destined, by its nature, to bring within the reach of the calculus the actual establishment of the differential equations; for, in a great number of important and difficult questions, such is the general effect of the *varied* equations, which, still more *indirect* than the simple differential equations with respect to the special objects of the investigation, are also much more easy to form, and from which we may then, by invariable and complete analytical methods, the object of which is to eliminate the new order of auxiliary infinitesimals which have been introduced, deduce those ordinary differential equations which it would often have been impossible to establish directly. The method of variations forms, then, the most sublime part of that vast system of mathematical analysis, which, setting out from the most simple elements of algebra, organizes, by an uninterrupted succession of ideas, general methods more and more powerful, for the study of natural philosophy, and which, in its whole, presents the most incomparably imposing and unequivocal monument of the power of the human intellect.

We must, however, also admit that the conceptions which are habitually considered in the method of variations being, by their nature, more indirect, more general, and especially more abstract than all others, the employment of such a method exacts necessarily and continuously the highest known degree of intellectual exertion, in order never to lose sight of the precise object of the investigation, in following reasonings which offer to the mind such uncertain resting-places, and in which signs are of scarcely any assistance. We must undoubtedly attribute in a great degree to this difficulty the little real use which geometers, with the exception of Lagrange, have as yet made of such an admirable conception.

CHAPTER VI.

THE CALCULUS OF FINITE DIFFERENCES.

THE different fundamental considerations indicated in the five preceding chapters constitute, in reality, all the essential bases of a complete exposition of mathematical analysis, regarded in the philosophical point of view. Nevertheless, in order not to neglect any truly important general conception relating to this analysis, I think that I should here very summarily explain the veritable character of a kind of calculus which is very extended, and which, though at bottom it really belongs to ordinary analysis, is still regarded as being of an essentially distinct nature. I refer to the *Calculus of Finite Differences*, which will be the special subject of this chapter.

Its general Character. This calculus, created by Taylor, in his celebrated work entitled *Methodus Incrementorum*, consists essentially in the consideration of the finite increments which functions receive as a consequence of analogous increments on the part of the corresponding variables. These increments or *differences*, which take the characteristic Δ, to distinguish them from *differentials*, or infinitely small increments, may be in their turn regarded as new functions, and become the subject of a second similar consideration, and so on; from which results the notion of differences of various successive orders, analogous, at least in appearance, to the consecutive orders of differentials. Such a calculus evidently presents, like the calculus of indirect functions, two general classes of questions:

1°. To determine the successive differences of all the various analytical functions of one or more variables, as the result of a definite manner of increase of the independent variables, which are generally supposed to augment in arithmetical progression:

2°. Reciprocally, to start from these differences, or, more generally, from any equations established between them, and go back to the primitive functions themselves, or to their corresponding relations.

Hence follows the decomposition of this calculus into two distinct ones, to which are usually given the names of the *Direct*, and the *Inverse Calculus of Finite Differences*, the latter being also sometimes called the *Integral Calculus of Finite Differences.* Each of these would, also, evidently admit of a logical distribution similar to that given in the fourth chapter for the differential and the integral calculus.

Its true Nature. There is no doubt that Taylor thought that by such a conception he had founded a calculus of an entirely new nature, absolutely distinct from ordinary analysis, and more general than the calculus of Leibnitz, although resting on an analogous consideration. It is in this way, also, that almost all geometers have viewed the analysis of Taylor; but Lagrange, with his usual profundity, clearly perceived that these properties belonged much more to the forms and to the notations employed by Taylor than to the substance of his theory. In fact, that which constitutes the peculiar character of the analysis of Leibnitz, and makes of it a truly distinct and superior calculus, is the circumstance that the derived functions are in general of an entirely different nature from the primitive functions, so that they may give rise to more simple and more easily formed relations; whence result the admirable fundamental properties of the transcendental analysis, which have been already explained. But it is not so with the *differences* considered by Taylor; for these differences are, by their nature, functions essentially similar to those which have produced them, a circumstance which renders them unsuitable to facilitate the establishment of equations, and prevents their leading to more general relations. Every equation of finite differences is truly, at bottom, an equation directly relating to the very magnitudes whose successive states are compared. The scaffolding of new signs, which produce an illusion respecting the true character of these equations, disguises it, however, in a very imperfect manner, since it could always be easily made apparent by replacing the *differences* by the equivalent combinations of the primitive magnitudes, of which they are really only the abridged designations. Thus the calculus of Taylor never has offered, and never can offer, in any question of geometry or of mechanics, that powerful general aid which we have seen to result necessarily from the analysis of Leibnitz. Lagrange has, moreover, very clearly proven that the pretended analogy observed between the calculus of differences and the infinitesimal calculus was radically vicious, in this way, that the formulas belonging to the former calculus can never furnish, as particular cases, those which belong to the latter, the nature of which is essentially distinct.

From these considerations I am led to think that the calculus of finite differences is, in general, improperly classed with the transcendental analysis proper, that is, with the calculus of indirect functions. I consider it, on the contrary, in accordance with the views of Lagrange, to be only a very extensive and very important branch of ordinary analysis, that is to say, of that which I have named the calculus of direct functions, the equations which it considers being always, in spite of the notation, simple *direct* equations.

GENERAL THEORY OF SERIES.

To sum up as briefly as possible the preceding explanation, the calculus of Taylor ought to be regarded as having constantly for its true object the general theory of *Series*, the most simple cases of which had alone been considered before that illustrious geometer. I ought, properly, to have mentioned this important theory in treating, in the second chapter, of Algebra proper, of which it is such an extensive branch. But, in order to avoid a double reference to it, I have preferred to notice it only in the consideration of the calculus of finite differences, which, reduced to its most simple general expression, is nothing but a complete logical study of questions relating to *series*.

Every *Series*, or succession of numbers deduced from one another according to any constant law, necessarily gives rise to these two fundamental questions:

1°. The law of the series being supposed known, to find the expression for its general term, so as to be able to calculate immediately any term whatever without being obliged to form successively all the preceding terms:

2°. In the same circumstances, to determine the *sum* of any number of terms of the series by means of their places, so that it can be known without the necessity of continually adding these terms together.

These two fundamental questions being considered to be resolved, it may be proposed, reciprocally, to find the law of a series from the form of its general term, or the expression of the sum. Each of these different problems has so much the more extent and difficulty, as there can be conceived a greater number of different *laws* for the series, according to the number of preceding terms on which each term directly depends, and according to the function which expresses that dependence. We may even consider series with several variable indices, as Laplace has done in his "Analytical Theory of Probabilities," by the analysis to which he has given the name of *Theory of generating Functions*, although it is really only a new and higher branch of the calculus of finite differences or of the general theory of series.

These general views which I have indicated give only an imperfect idea of the truly infinite extent and variety of the questions to which geometers have risen by means of this single consideration of series, so simple in appearance and so limited in its origin. It necessarily presents as many different cases as the algebraic resolution of equations, considered in its whole extent; and it is, by its nature, much more complicated, so much, indeed, that it always needs this last to conduct it to a complete solution. We may, therefore, anticipate what must still be its extreme imperfection, in spite of the successive labours of several geometers of the first order. We do not, indeed, possess as yet the complete and logical solution of any but the most simple questions of this nature.

Its identity with this Calculus. It is now easy to conceive the necessary and perfect identity, which has been already announced, between the calculus of finite differences and the theory of series considered in all its bearings. In fact, every differentiation after the manner of Taylor evidently amounts to finding the *law* of formation of a series with one or with several variable indices, from the expression of its general term; in the same way, every analogous integration may be regarded as having for its object the summation of a series, the general term of which would be expressed by the proposed difference. In this point of view, the various problems of the calculus of differences, direct or inverse, resolved by Taylor and his successors, have really a very great value, as treating of important questions relating to series. But it is very doubtful if the form and the notation introduced by Taylor really give any essential facility in the solution of questions of this kind. It would be, perhaps, more advantageous for most cases, and certainly more logical, to replace the *differences* by the terms themselves, certain combinations of which they represent. As the calculus of Taylor does not rest on a truly distinct fundamental idea, and has nothing peculiar to it but its system of signs, there could never really be any important advantage in considering it as detached from ordinary analysis, of which it is, in reality, only an immense branch. This consideration of *differences*, most generally useless, even if it does not cause complication, seems to me to retain the character of an epoch in which, analytical ideas not being sufficiently familiar to geometers, they were naturally led to prefer the special forms suitable for simple numerical comparisons.

PERIODIC OR DISCONTINUOUS FUNCTIONS.

However that may be, I must not finish this general appreciation of the calculus of finite differences without noticing a new conception to which it has given birth, and which has since acquired a great importance. It is the consideration of those periodic or discontinuous functions which preserve the same value for an infinite series of values of the corresponding variables, subjected to a certain law, and which must be necessarily added to the integrals of the equations of finite differences in order to render them sufficiently general, as simple arbitrary constants are added to all quadratures in order to complete their generality. This idea, primitively introduced by Euler, has since been the subject of extended investigation by M. Fourier, who has made new and important applications of it in his mathematical theory of heat.

APPLICATIONS OF THIS CALCULUS.

Series. Among the principal general applications which have been made of the calculus of finite differences, it would be proper to place in the first rank, as the most extended and the most important, the solution of questions relating to series; if, as has been shown, the general theory of series ought not to be considered as constituting, by its nature, the actual foundation of the calculus of Taylor.

Interpolations. This great class of problems being then set aside, the most essential of the veritable applications of the analysis of Taylor is, undoubtedly, thus far, the general method of *interpolations*, so frequently and so usefully employed in the investigation of the empirical laws of natural phenomena. The question consists, as is well known, in intercalating between certain given numbers other intermediate numbers, subjected to the same law which we suppose to exist between the first. We can abundantly verify, in this principal application of the calculus of Taylor, how truly foreign and often inconvenient is the consideration of *differences* with respect to the questions which depend on that analysis. Indeed, Lagrange has replaced the formulas of interpolation, deduced from the ordinary algorithm of the calculus of finite differences, by much simpler general formulas, which are now almost always preferred, and which have been found directly, without making any use of the notion of *differences*, which only complicates the question.

Approximate Rectification, &c. A last important class of applications of the calculus of finite differences, which deserves to be distinguished from the preceding, consists in the eminently useful employment made of it in geometry for determining by approximation the length and the area of any curve, and in the same way the cubature of a body of any form whatever. This procedure (which may besides be conceived abstractly as depending on the same analytical investigation as the question of interpolation) frequently offers a valuable supplement to the entirely logical geometrical methods which often lead to integrations, which we do not yet know how to effect, or to calculations of very complicated execution.

Such are the various principal considerations to be noticed with respect to the calculus of finite differences. This examination completes the proposed philosophical outline of ABSTRACT MATHEMATICS.

Concrete Mathematics will now be the subject of a similar labour. In it we shall particularly devote ourselves to examining how it has been possible (supposing the general science of the calculus to be perfect), by invariable procedures, to reduce to pure questions of analysis all the problems which can be presented by *Geometry* and *Mechanics*, and thus to impress on these two fundamental bases of natural philosophy a degree of precision and especially of unity; in a word, a character of high perfection, which could be communicated to them by such a course alone.

BOOK II.

GEOMETRY.

BOOK II.

GEOMETRY.

CHAPTER I.

GENERAL VIEW OF GEOMETRY.

Its true Nature. After the general exposition of the philosophical character of concrete mathematics, compared with that of abstract mathematics, given in the introductory chapter, it need not here be shown in a special manner that geometry must be considered as a true natural science, only much more simple, and therefore much more perfect, than any other. This necessary perfection of geometry, obtained essentially by the application of mathematical analysis, which it so eminently admits, is apt to produce erroneous views of the real nature of this fundamental science, which most minds at present conceive to be a purely logical science quite independent of observation. It is nevertheless evident, to any one who examines with attention the character of geometrical reasonings, even in the present state of abstract geometry, that, although the facts which are considered in it are much more closely united than those relating to any other science, still there always exists, with respect to every body studied by geometers, a certain number of primitive phenomena, which, since they are not established by any reasoning, must be founded on observation alone, and which form the necessary basis of all the deductions.

The scientific superiority of geometry arises from the phenomena which it considers being necessarily the most universal and the most simple of all. Not only may all the bodies of nature give rise to geometrical inquiries, as well as mechanical ones, but still farther; geometrical phenomena would still exist, even though all the parts of the universe should be considered as immovable. Geometry is then, by its nature, more general than mechanics. At the same time, its phenomena are more simple, for they are evidently independent of mechanical phenomena, while these latter are always complicated with the former. The same relations hold good in comparing geometry with abstract thermology.

For these reasons, in our classification we have made geometry the first part of concrete mathematics; that part the study of which, in addition to its own importance, serves as the indispensable basis of all the rest.

Before considering directly the philosophical study of the different orders of inquiries which constitute our present geometry, we should obtain a clear and exact idea of the general destination of that science, viewed in all its bearings. Such is the object of this chapter.

Definition. Geometry is commonly defined in a very vague and entirely improper manner, as being *the science of extension.* An improvement on this would be to say that geometry has for its object the *measurement* of extension; but such an explanation would be very insufficient, although at bottom correct, and would be far from giving any idea of the true general character of geometrical science.

To do this, I think that I should first explain *two fundamental ideas*, which, very simple in themselves, have been singularly obscured by the employment of metaphysical considerations.

The Idea of Space. The first is that of *Space*. This conception properly consists simply in this, that, instead of considering extension in the bodies themselves, we view it in an indefinite medium, which we regard as containing all the bodies of the universe. This notion is naturally suggested to us by observation, when we think of the *impression* which a body would leave in a fluid in which it had been placed. It is clear, in fact, that, as regards its geometrical relations, such an *impression* may be substituted for the body itself, without altering the reasonings respecting it. As to the physical nature of this indefinite *space*, we are spontaneously led to represent it to ourselves, as being entirely analogous to the actual medium in which we live; so that if this medium was liquid instead of gaseous, our geometrical *space* would undoubtedly be conceived as liquid also. This circumstance is, moreover, only very secondary, the essential object of such a conception being only to make us view extension separately from the bodies which manifest it to us. We can easily understand in advance the importance of this fundamental image, since it permits us to study geometrical phenomena in themselves, abstraction being made of all the other phenomena which constantly accompany them in real bodies, without, however, exerting any influence over them. The regular establishment of this general abstraction must be regarded as the first step which has been made in the rational study of geometry, which would have been impossible if it had been necessary to consider, together with the form and the magnitude of bodies, all their other physical properties. The use of such an hypothesis, which is perhaps the most ancient philosophical conception created by the human mind, has now become so familiar to us, that we have difficulty in exactly estimating its importance, by trying to appreciate the consequences which would result from its suppression.

Different Kinds of Extension. The second preliminary geometrical conception which we have to examine is that of the different kinds of extension, designated by the words *volume*, *surface*, *line*, and even *point*, and of which the ordinary explanation is so unsatisfactory.*

Although it is evidently impossible to conceive any extension absolutely deprived of any one of the three fundamental dimensions, it is no less incontestable that, in a great number of occasions, even of immediate utility, geometrical questions depend on only two dimensions, considered separately from the third, or on a single dimension, considered separately from the two others. Again, independently of this direct motive, the study of extension with a single dimension, and afterwards with two, clearly presents itself as an indispensable preliminary for facilitating the study of complete bodies of three dimensions, the immediate theory of which would be too com-

* Lacroix has justly criticised the expression of *solid*, commonly used by geometers to designate a *volume*. It is certain, in fact, that when we wish to consider separately a certain portion of indefinite space, conceived as gaseous, we mentally solidify its exterior envelope, so that a *line* and a *surface* are habitually, to our minds, just as *solid* as a *volume*. It may also be remarked that most generally, in order that bodies may penetrate one another with more facility, we are obliged to imagine the interior of the *volumes* to be hollow, which renders still more sensible the impropriety of the word *solid*.

plicated. Such are the two general motives which oblige geometers to consider separately extension with regard to one or to two dimensions, as well as relatively to all three together.

The general notions of *surface* and of *line* have been formed by the human mind, in order that it may be able to think, in a permanent manner, of extension in two directions, or in one only. The hyperbolical expressions habitually employed by geometers to define these notions tend to convey false ideas of them; but, examined in themselves, they have no other object than to permit us to reason with facility respecting these two kinds of extension, making complete abstraction of that which ought not to be taken into consideration. Now for this it is sufficient to conceive the dimension which we wish to eliminate as becoming gradually smaller and smaller, the two others remaining the same, until it arrives at such a degree of tenuity that it can no longer fix the attention. It is thus that we naturally acquire the real idea of a *surface*, and, by a second analogous operation, the idea of a *line*, by repeating for breadth what we had at first done for thickness. Finally, if we again repeat the same operation, we arrive at the idea of a *point*, or of an extension considered only with reference to its place, abstraction being made of all magnitude, and designed consequently to determine positions.

Surfaces evidently have, moreover, the general property of exactly circumscribing volumes; and in the same way, *lines*, in their turn, circumscribe *surfaces* and are limited by *points*. But this consideration, to which too much importance is often given, is only a secondary one.

Surfaces and lines are, then, in reality, always conceived with three dimensions; it would be, in fact, impossible to represent to one's self a surface otherwise than as an extremely thin plate, and a line otherwise than as an infinitely fine thread. It is even plain that the degree of tenuity attributed by each individual to the dimensions of which he wishes to make abstraction is not constantly identical, for it must depend on the degree of subtilty of his habitual geometrical observations. This want of uniformity has, besides, no real inconvenience, since it is sufficient, in order that the ideas of surface and of line should satisfy the essential condition of their destination, for each one to represent to himself the dimensions which are to be neglected as being smaller than all those whose magnitude his daily experience gives him occasion to appreciate.

We hence see how devoid of all meaning are the fantastic discussions of metaphysicians upon the foundations of geometry. It should also be remarked that these primordial ideas are habitually presented by geometers in an unphilosophical manner, since, for example, they explain the notions of the different sorts of extent in an order absolutely the inverse of their natural dependence, which often produces the most serious inconveniences in elementary instruction.

THE FINAL OBJECT OF GEOMETRY.

These preliminaries being established, we can proceed directly to the general definition of geometry, continuing to conceive this science as having for its final object the *measurement* of extension.

It is necessary in this matter to go into a thorough

explanation, founded on the distinction of the three kinds of extension, since the notion of *measurement* is not exactly the same with reference to surfaces and volumes as to lines.

Nature of Geometrical Measurement. If we take the word *measurement* in its direct and general mathematical acceptation, which signifies simply the determination of the value of the *ratios* between any homogeneous magnitudes, we must consider, in geometry, that the *measurement* of surfaces and of volumes, unlike that of lines, is never conceived, even in the most simple and the most favourable cases, as being effected directly. The comparison of two lines is regarded as direct; that of two surfaces or of two volumes is, on the contrary, always indirect. Thus we conceive that two lines may be superposed; but the superposition of two surfaces, or, still more so, of two volumes, is evidently impossible in most cases; and, even when it becomes rigorously practicable, such a comparison is never either convenient or exact. It is, then, very necessary to explain wherein properly consists the truly geometrical measurement of a surface or of a volume.

Measurement of Surfaces and of Volumes. For this we must consider that, whatever may be the form of a body, there always exists a certain number of lines, more or less easy to be assigned, the length of which is sufficient to define exactly the magnitude of its surface or of its volume. Geometry, regarding these lines as alone susceptible of being directly measured, proposes to deduce, from the simple determination of them, the ratio of the surface or of the volume sought, to the unity of surface, or to the unity of volume. Thus the general object of geometry, with respect to surfaces and to volumes, is properly to reduce all comparisons of surfaces or of volumes to simple comparisons of lines.

Besides the very great facility which such a transformation evidently offers for the measurement of volumes and of surfaces, there results from it, in considering it in a more extended and more scientific manner, the general possibility of reducing to questions of lines all questions relating to volumes and to surfaces, considered with reference to their magnitude. Such is often the most important use of the geometrical expressions which determine surfaces and volumes in functions of the corresponding lines.

It is true that direct comparisons between surfaces or between volumes are sometimes employed; but such measurements are not regarded as geometrical, but only as a supplement sometimes necessary, although too rarely applicable, to the insufficiency or to the difficulty of truly rational methods. It is thus that we often determine the volume of a body, and in certain cases its surface, by means of its weight. In the same way, on other occasions, when we can substitute for the proposed volume an equivalent liquid volume, we establish directly the comparison of the two volumes, by profiting by the property possessed by liquid masses, of assuming any desired form. But all means of this nature are purely mechanical, and rational geometry necessarily rejects them.

To render more sensible the difference between these modes of determination and true geometrical measurements, I will cite a single very remarkable example; the manner in which Galileo determined the ratio of the ordinary cycloid to that of the generating circle. The geometry of his time was as yet insufficient for the rational solution of such a problem. Galileo conceived the idea of discovering that ratio by a direct experiment. Having weighed as exactly as possible two plates of the same material and of equal thickness, one of them having the form of a circle and the other that of the generated cycloid, he found the weight of the latter always triple that of the former; whence he inferred that the area of the cycloid is triple that of the generating circle, a result agreeing with the veritable solution subsequently obtained by Pascal and Wallis. Such a success evidently depends on the extreme simplicity of the ratio sought; and we can understand the necessary insufficiency of such expedients, even when they are actually practicable.

We see clearly, from what precedes, the nature of that part of geometry relating to *volumes* and that relating to *surfaces*. But the character of the geometry of *lines* is not so apparent, since, in order to simplify the exposition, we have considered the measurement of lines as being made directly. There is, therefore, needed a complementary explanation with respect to them.

Measurement of curved Lines. For this purpose, it is sufficient to distinguish between the right line and curved lines, the measurement of the first being alone regarded as direct, and that of the other as always indirect. Although superposition is sometimes strictly practicable for curved lines, it is nevertheless evident that truly rational geometry must necessarily reject it, as not admitting of any precision, even when it is possible. The geometry of lines has, then, for its general object, to reduce in every case the measurement of curved lines to that of right lines; and consequently, in the most extended point of view, to reduce to simple questions of right lines all questions relating to the magnitude of any curves whatever. To understand the possibility of such a transformation, we must remark, that in every curve there always exist certain right lines, the length of which must be sufficient to determine that of the curve. Thus, in a circle, it is evident that from the length of the radius we must be able to deduce that of the circumference; in the same way, the length of an ellipse depends on that of its two axes; the length of a cycloid upon the diameter of the generating circle, &c.; and if, instead of considering the whole of each curve, we demand, more generally, the length of any arc, it will be sufficient to add to the different rectilinear parameters, which determine the whole curve, the chord of the proposed arc, or the co-ordinates of its extremities. To discover the relation which exists between the length of a curved line and that of similar right lines, is the general problem of the part of geometry which relates to the study of lines.

Combining this consideration with those previously suggested with respect to volumes and to surfaces, we may form a very clear idea of the science of geometry, conceived in all its parts, by assigning to it, for its general object, the final reduction of the comparisons of all kinds of extent, volumes, surfaces, or lines, to simple comparisons of right lines, the only comparisons regarded as capable of being made directly, and which indeed could not be reduced to any others more easy to effect. Such a conception, at the same time, indicates clearly the veritable character of geometry, and seems suited to show at a single glance its utility and its perfection.

Measurement of right Lines. In order to complete this fundamental explanation, I have yet to show how there can be, in geometry, a special section relating to the right line, which seems at first incompatible with the principle that the measurement of this class of lines must always be regarded as direct.

It is so, in fact, as compared with that of curved lines, and of all the other objects which geometry considers. But it is evident that the estimation of a right line cannot be viewed as direct except so far as the linear unit can be applied to it. Now this often presents insurmountable difficulties, as I had occasion to show, for another reason, in the introductory chapter. We must, then, make the measurement of the proposed right line depend on other analogous measurements capable of being effected directly. There is, then, necessarily a primary distinct branch of geometry, exclusively devoted to the right line; its object is to determine certain right lines from others by means of the relations belonging to the figures resulting from their assemblage. This preliminary part of geometry, which is almost imperceptible in viewing the whole of the science, is nevertheless susceptible of a great development. It is evidently of especial importance, since all other geometrical measurements are referred to those of right lines, and if they could not be determined, the solution of every question would remain unfinished.

Such, then, are the various fundamental parts of rational geometry, arranged according to their natural dependence; the geometry of *lines* being first considered, beginning with the right line; then the geometry of *surfaces*, and, finally, that of *solids*.

INFINITE EXTENT OF ITS FIELD.

Having determined with precision the general and final object of geometrical inquiries, the science must now be considered with respect to the field embraced by each of its three fundamental sections.

Thus considered, geometry is evidently susceptible, by its nature, of an extension which is rigorously infinite; for the measurement of lines, of surfaces, or of volumes presents necessarily as many distinct questions as we can conceive different figures subjected to exact definitions; and their number is evidently infinite.

Geometers limited themselves at first to consider the most simple figures which were directly furnished them by nature, or which were deduced from these primitive elements by the least complicated combinations. But they have perceived, since Descartes, that, in order to constitute the science in the most philosophical manner, it was necessary to make it apply to all imaginable figures. This abstract geometry will then inevitably comprehend as particular cases all the different real figures which the exterior world could present. It is then a fundamental principle in truly rational geometry to consider, as far as possible, all figures which can be rigorously conceived.

The most superficial examination is enough to convince us that these figures present a variety which is quite infinite.

Infinity of Lines. With respect to curved *lines*, regarding them as generated by the motion of a point governed by a certain law, it is plain that we shall have, in general, as many different curves as we conceive different laws for this motion, which may evidently be determined by an infinity of distinct conditions; although it may sometimes accidentally happen that new generations produce curves which have been already obtained. Thus, among plane curves, if a point moves so as to remain constantly at the same distance from a fixed point, it will generate a *circle;* if it is the sum or the difference of its distances from two fixed points which remains constant, the curve described will be an *ellipse* or an *hyperbola;* if it is their product, we shall have an entirely different curve; if the point departs equally from a fixed point and from a fixed line, it will describe a *parabola;* if it revolves on a circle at the same time that this circle rolls along a straight line, we shall have a *cycloid;* if it advances along a straight line, while this line, fixed at one of its extremities, turns in any manner whatever, there will result what in general terms are called *spirals,* which of themselves evidently present as many perfectly distinct curves as we can suppose different relations between these two motions of translation and of rotation, &c. Each of these different curves may then furnish new ones, by the different general constructions which geometers have imagined, and which give rise to evolutes, to epicycloids, to caustics, &c. Finally, there exists a still greater variety among curves of double curvature.

Infinity of Surfaces. As to *surfaces*, the figures are necessarily more different still, considering them as generated by the motion of lines. Indeed, the figure may then vary, not only in considering, as in curves, the different infinitely numerous laws to which the motion of the generating line may be subjected, but also in supposing that this line itself may change its nature; a circumstance which has nothing analogous in curves, since the points which describe them cannot have any distinct figure. Two classes of very different conditions may then cause the figures of surfaces to vary, while there exists only one for lines. It is useless to cite examples of this doubly infinite multiplicity of surfaces. It would be sufficient to consider the extreme variety of the single group of surfaces which may be generated by a right line, and which comprehends the whole family of cylindrical surfaces, that of conical surfaces, the most general class of developable surfaces, &c.

Infinity of Volumes. With respect to *volumes*, there is no occasion for any special consideration, since they are distinguished from each other only by the surfaces which bound them.

In order to complete this sketch, it should be added that surfaces themselves furnish a new general means of conceiving new curves, since every curve may be regarded as produced by the intersection of two surfaces. It is in this way, indeed, that the first lines which we may regard as having been truly invented by geometers were obtained, since nature gave directly the straight line and the circle. We know that the ellipse, the parabola, and the hyperbola, the only curves completely studied by the ancients, were in their origin conceived only as resulting from the intersection of a cone with circular base by a plane in different positions. It is evident that, by the combined employment of these different general means for the formation of lines and of surfaces, we could produce a rigorously infinitely series of distinct forms in starting from only a very small number of figures directly furnished by observation.

Analytical invention of Curves, &c. Finally, all the various direct means for the invention of figures have scarcely any farther importance, since rational geometry has assumed its final character in the hands of Descartes. Indeed, as we shall see more fully in chapter iii., the invention of figures is now reduced to the invention of equations, so that nothing is more easy than to conceive new lines and new surfaces, by changing at will the functions introduced into the equations. This simple abstract procedure is, in this respect, infinitely more fruitful than all the direct resources of geometry, developed by the most powerful imagination, which should devote itself exclusively to that order of conceptions. It also explains, in the most general and the most striking manner, the necessarily infinite variety of geometrical forms, which thus corresponds to the diversity of analytical functions. Lastly, it shows no less clearly that the different forms of surfaces must be still more numerous than those of lines, since lines are represented analytically by equations with two variables, while surfaces give rise to equations with three variables, which necessarily present a greater diversity.

The preceding considerations are sufficient to show clearly the rigorously infinite extent of each of the three general sections of geometry.

EXPANSION OF ORIGINAL DEFINITION.

To complete the formation of an exact and sufficiently extended idea of the nature of geometrical inquiries, it is now indispensable to return to the general definition above given, in order to present it under a new point of view, without which the complete science would be only very imperfectly conceived.

When we assign as the object of geometry the *measurement* of all sorts of lines, surfaces, and volumes, that is, as has been explained, the reduction of all geometrical comparisons to simple comparisons of right lines, we have evidently the advantage of indicating a general destination very precise and very easy to comprehend. But if we set aside every definition, and examine the actual composition of the science of geometry, we will at first be induced to regard the preceding definition as much too narrow; for it is certain that the greater part of the investigations which constitute our present geometry do not at all appear to have for their object the *measurement* of extension. In spite of this fundamental objection, I will persist in retaining this definition; for, in fact, if, instead of confining ourselves to considering the different questions of geometry isolatedly, we endeavour to grasp the leading questions, in comparison with which all others, however important they may be, must be regarded as only secondary, we will finally recognize that the measurement of lines, of surfaces, and of volumes, is the invariable object, sometimes *direct*, though most often *indirect*, of all geometrical labours.

This general proposition being fundamental, since it can alone give our definition all its value, it is indispensable to enter into some developments upon this subject.

PROPERTIES OF LINES AND SURFACES.

When we examine with attention the geometrical investigations which do not seem to relate to the *measurement* of extent, we find that they consist essentially in the study of the different *properties* of each line or of each surface; that is, in the knowledge of the different modes of generation, or at least of definition, peculiar to each figure considered. Now we can easily establish in the most general manner the necessary relation of such a study to the question of *measurement*, for which the most complete knowledge of the properties of each form is an indispensable preliminary. This is concurrently proven by two considerations, equally fundamental, although quite distinct in their nature.

NECESSITY OF THEIR STUDY: 1. *To find the most suitable Property.* The *first*, purely scientific, consists in remarking that, if we did not know any other characteristic property of each line or surface than that one according to which geometers had first conceived it, in most cases it would be impossible to succeed in the solution of questions relating to its *measurement*. In fact, it is easy to understand that the different definitions which each figure admits of are not all equally suitable for such an object, and that they even present the most complete oppositions in that respect. Besides, since the primitive definition of each figure was evidently not chosen with this condition in view, it is clear that we must not expect, in general, to find it the most suitable; whence results the necessity of discovering others, that is, of studying as far as is possible the *properties* of the proposed figure. Let us suppose, for example, that the circle is defined to be "the curve which, with the same contour, contains the greatest area." This is certainly a very characteristic property, but we would evidently find insurmountable difficulties in trying to deduce from such a starting point the solution of the fundamental questions relating to the rectification or to the quadrature of this curve. It is clear, in advance, that the property of having all its points equally distant from a fixed point must evidently be much better adapted to inquiries of this nature, even though it be not precisely the most suitable. In like manner, would Archimedes ever have been able to discover the quadrature of the parabola if he had known no other property of that curve than that it was the section of a cone with a circular base, by a plane parallel to its generatrix? The purely speculative labours of preceding geometers, in transforming this first definition, were evidently indispensable preliminaries to the direct solution of such a question. The same is true, in a still greater degree, with respect to surfaces. To form a just idea of this, we need only compare, as to the question of cubature or quadrature, the common definition of the sphere with that one, no less characteristic certainly, which would consist in regarding a spherical body, as that one which, with the same area, contains the greatest volume.

No more examples are needed to show the necessity of knowing, so far as is possible, all the properties of each line or of each surface, in order to facilitate the investigation of rectifications, of quadratures, and of cubatures, which constitutes the final object of geometry. We may even say that the principal difficulty of questions of this kind consists in employing in each case the property which is best adapted to the nature of the proposed problem. Thus, while we continue to indicate, for more precision, the measurement of extension as the general destination of geometry, this first consideration, which goes to the very bottom of the subject, shows clearly the necessity of including in it the study, as thorough as possible, of the different generations or definitions belonging to the same form.

2. *To pass from the Concrete to the Abstract.* A *second* consideration, of at least equal importance, consists in such a study being indispensable for organizing in a rational manner the relation of the abstract to the concrete in geometry.

The science of geometry having to consider all imaginable figures which admit of an exact definition, it necessarily results from this, as we have remarked, that questions relating to any figures presented by nature are always implicitly comprised in this abstract geometry, supposed to have attained its perfection. But when it is necessary to actually pass to concrete geometry, we constantly meet with a fundamental difficulty, that of knowing to which of the different abstract types we are to refer, with sufficient approximation, the real lines or surfaces which we have to study. Now it is for the purpose of establishing such a relation that it is particularly indispensable to know the greatest possible number of properties of each figure considered in geometry.

In fact, if we always confined ourselves to the single primitive definition of a line or of a surface, supposing even that we could then *measure* it (which, according to the first order of considerations, would generally be impossible), this knowledge would remain almost necessarily barren in the application, since we should not ordinarily know how to recognize that figure in nature when it presented itself there; to ensure that, it would be necessary that the single characteristic, according to which geometers had conceived it, should be precisely that one whose verification external circumstances would admit: a coincidence which would be purely fortuitous, and on which we could not count, although it might sometimes take place. It is, then, only by multiplying as much as possible the characteristic properties of each abstract figure, that we can be assured, in advance, of recognizing it in the concrete state, and of thus turning to account all our rational labours, by verifying in each case the definition which is susceptible of being directly proven. This definition is almost always the only one in given circumstances, and varies, on the other hand, for the same figure, with different circumstances; a double reason for its previous determination.

Illustration: Orbits of the Planets. The geometry of the heavens furnishes us with a very memorable example in this matter, well suited to show the general necessity of such a study. We know that the ellipse was discovered by Kepler to be the curve which the planets describe about the sun, and the satellites about their planets. Now would this fundamental discovery, which re-created astronomy, ever have been possible, if geometers had been always confined to conceiving the ellipse only as the oblique section of a circular cone by a plane? No such definition, it is evident, would admit of such a verification. The most general property of the ellipse, that the sum of the distances from any of its points to two fixed points is a constant quantity, is undoubtedly much more susceptible, by its nature, of causing the curve to be recognized in this case, but still is not directly suitable. The only characteristic which can here be immediately verified is that which is derived from the relation which exists in the ellipse between the length of the focal distances and their direction; the only relation which admits of an astronomical interpretation, as expressing the law which connects the distance from the planet to the sun, with the time elapsed since the beginning of its revolution. It was, then, necessary that the purely speculative labours of the Greek geometers on the properties of the conic sections should have previously presented their generation under a multitude of different points of view, before Kepler could thus pass from the abstract to the concrete, in choosing from among all these different characteristics that one which could be most easily proven for the planetary orbits.

Illustration: Figure of the Earth. Another example of the same order, but relating to surfaces, occurs in considering the important question of the figure of the earth. If we had never known any other property of the sphere than its primitive character of having all its points equally distant from an interior point, how would we ever have been able to discover that the surface of the earth was spherical? For this, it was necessary previously to deduce from this definition of the sphere some properties capable of being verified by observations made upon the surface alone, such as the constant ratio which exists between the length of the path traversed in the direction of any meridian of a sphere going towards a pole, and the angular height of this pole above the horizon at each point. Another example, but involving a much longer series of preliminary speculations, is the subsequent proof that the earth is not rigorously spherical, but that its form is that of an ellipsoid of revolution.

After such examples, it would be needless to give any others, which any one besides may easily multiply. All of them prove that, without a very extended knowledge of the different properties of each figure, the relation of the abstract to the concrete, in geometry, would be purely accidental, and that the science would consequently want one of its most essential foundations.

Such, then, are two general considerations which fully demonstrate the necessity of introducing into geometry a great number of investigations which have not the *measurement* of extension for their direct object; while we continue, however, to conceive such a measurement as being the final destination of all geometrical science. In this way we can retain the philosophical advantages of the clearness and precision of this definition, and still include in it, in a very logical though indirect manner, all known geometrical researches, in considering those which do not seem to relate to the measurement of extension, as intended either to prepare for the solution of the final questions, or to render possible the application of the solutions obtained.

Having thus recognized, as a general principle, the close and necessary connexion of the study of the properties of lines and surfaces with those researches which constitute the final object of geometry, it is evident that geometers, in the progress of their labours, must by no means constrain themselves to keep such a connexion always in view. Knowing, once for all, how important it is to vary as much as possible the manner of conceiving each figure, they should pursue that study, without considering of what immediate use such or such a special property may be for rectifications, quadratures, and cubatures. They would uselessly fetter their inquiries by attaching a puerile importance to the continued establishment of that co-ordination.

This general exposition of the general object of geometry is so much the more indispensable, since, by the very nature of the subject, this study of the different properties of each line and of each surface necessarily composes by far the greater part of the whole body of geometrical researches. Indeed, the questions directly relating to rectifications, to quadratures, and to cubatures, are evidently, by themselves, very few in number for each figure considered. On the other hand, the study of the properties of the same figure presents an unlimited field to the activity of the human mind, in which it may always hope to make new discoveries. Thus, although geometers have occupied themselves for twenty centuries, with more or less activity undoubtedly, but without any real interruption, in the study of the conic sections, they are far from regarding that so simple subject as being exhausted ; and it is certain, indeed, that in continuing to devote themselves to it, they would not fail to find still unknown properties of those different curves. If labours of this kind have slackened considerably for a century past, it is not because they are completed, but only, as will be presently explained, because the philosophical revolution in geometry, brought about by Descartes, has singularly diminished the importance of such researches.

It results from the preceding considerations that not only is the field of geometry necessarily infinite because of the variety of figures to be considered, but also in virtue of the diversity of the points of view under the same figure may be regarded. This last conception is, indeed, that which gives the broadest and most complete idea of the whole body of geometrical researches. We see that studies of this kind consist essentially, for each line or for each surface, in connecting all the geometrical phenomena which it can present, with a single fundamental phenomenon, regarded as the primitive definition.

THE TWO GENERAL METHODS OF GEOMETRY.

Having now explained in a general and yet precise manner the final object of geometry, and shown how the science, thus defined, comprehends a very extensive class of researches which did not at first appear necessarily to belong to it, there remains to be considered the *method* to be followed for the formation of this science. This discussion is indispensable to complete this first sketch of the philosophical character of geometry. I shall here confine myself to indicating the most general consideration in this matter, developing and summing up this important fundamental idea in the following chapters.

Geometrical questions may be treated according to *two methods* so different, that there result from them two sorts of geometry, so to say, the philosophical character of which does not seem to me to have yet been properly apprehended. The expressions of *Synthetical Geometry* and *Analytical Geometry*, habitually employed to designate them, give a very false idea of them. I would much prefer the purely historical denominations of *Geometry of the Ancients* and *Geometry of the Moderns*, which have at least the advantage of not causing their true character to be misunderstood. But I propose to employ henceforth the regular expressions of *Special Geometry* and *General Geometry*, which seem to me suited to characterize with precision the veritable nature of the two methods.

Their fundamental Difference. The fundamental difference between the manner in which we conceive Geometry since Descartes, and the manner in which the geometers of antiquity treated geometrical questions, is not the use of the Calculus (or Algebra), as is commonly thought to be the case. On the one hand, it is certain that the use of the calculus was not entirely unknown to the ancient geometers, since they used to make continual and very extensive applications of the theory of proportions, which was for them, as a means of deduction, a sort of real, though very imperfect and especially extremely limited equivalent for our present algebra. The calculus may even be employed in a much more complete manner than they have used it, in order to obtain certain geometrical solutions, which will still retain all the essential character of the ancient geometry; this occurs very frequently with respect to those problems of geometry of two or of three dimensions, which are commonly designated under the name of *determinate*. On the other hand, important as is the influence of the calculus in our modern geometry, various solutions obtained without algebra may sometimes manifest the peculiar character which distinguishes it from the ancient geometry, although analysis is generally indispensable. I will cite, as an example, the method of Roberval for tangents, the nature of which is essentially modern, and which, however, leads in certain cases to complete solutions, without any aid from the calculus. It is not, then, the instrument of deduction employed which is the principal distinction between the two courses which the human mind can take in geometry.

The real fundamental difference, as yet imperfectly apprehended, seems to me to consist in the very nature of the questions considered. In truth, geometry, viewed as a whole, and supposed to have attained entire perfection, must, as we have seen on the one hand, embrace all imaginable figures, and, on the other, discover all the properties of each figure. It admits, from this double consideration, of being treated according to two essentially distinct plans; either, 1°, by grouping together all the questions, however different they may be, which relate to the same figure, and isolating those relating to different bodies, whatever analogy there may exist between them; or, 2°, on the contrary, by uniting under one point of view all similar inquiries, to whatever different figures they may relate, and separating the questions relating to the really different properties of the same body. In a word, the whole body of geometry may be essentially arranged either with reference to the *bodies* studied or to the *phenomena* to be considered. The first plan, which is the most natural, was that of the ancients; the second, infinitely more rational, is that of the moderns since Descartes.

Geometry of the Ancients. Indeed, the principal characteristics of the ancient geometry is that they studied, one by one, the different lines and the different surfaces, not passing to the examination of a new figure till they thought they had exhausted all that there was interesting in the figures already known. In this way of proceeding, when they undertook the study of a new curve, the whole of the labour bestowed on the preceding ones could not offer directly any essential assistance, otherwise than by the geometrical practice to which it had trained the mind. Whatever might be the real similarity of the questions proposed as to two different figures, the complete knowledge acquired for the one could not at all dispense with taking up again the whole of the investigation for the other. Thus the progress of the mind was never assured; so that they could not be certain, in advance, of obtaining any solution whatever, however analogous the proposed problem might be to questions which had been already resolved. Thus, for example, the determination of the tangents to the three conic sections did not furnish any rational assistance for drawing the tangent to any other new curve, such as the conchoid, the cissoid, &c. In a word, the geometry of the ancients was, according to the expression proposed above, essentially *special*.

Geometry of the Moderns. In the system of the moderns, geometry is, on the contrary, eminently *general*, that is to say, relating to any figures whatever. It is easy to understand, in the first place, that all geometrical expressions of any interest may be proposed with reference to all imaginable figures. This is seen directly in the fundamental problems—of rectifications, quadratures, and cubatures—which constitute, as has been shown, the final object of geometry. But this remark is no less incontestable, even for investigations which relate to the different *properties* of lines and of surfaces, and of which the most essential, such as the question of tangents or of tangent planes, the theory of curvatures,

&c., are evidently common to all figures whatever. The very few investigations which are truly peculiar to particular figures have only an extremely secondary importance. This being understood, modern geometry consists essentially in abstracting, in order to treat it by itself, in an entirely general manner, every question relating to the same geometrical phenomenon, in whatever bodies it may be considered. The application of the universal theories thus constructed to the special determination of the phenomenon which is treated of in each particular body, is now regarded as only a subaltern labour, to be executed according to invariable rules, and the success of which is certain in advance. This labour is, in a word, of the same character as the numerical calculation of an analytical formula. There can be no other merit in it than that of presenting in each case the solution which is necessarily furnished by the general method, with all the simplicity and elegance which the line or the surface considered can admit of. But no real importance is attached to any thing but the conception and the complete solution of a new question belonging to any figure whatever. Labours of this kind are alone regarded as producing any real advance in science. The attention of geometers, thus relieved from the examination of the peculiarities of different figures, and wholly directed towards general questions, has been thereby able to elevate itself to the consideration of new geometrical conceptions, which, applied to the curves studied by the ancients, have led to the discovery of important properties which they had not before even suspected. Such is geometry, since the radical revolution produced by Descartes in the general system of the science.

The Superiority of the modern Geometry. The mere indication of the fundamental character of each of the two geometries is undoubtedly sufficient to make apparent the immense necessary superiority of modern geometry. We may even say that, before the great conception of Descartes, rational geometry was not truly constituted upon definitive bases, whether in its abstract or concrete relations. In fact, as regards science, considered speculatively, it is clear that, in continuing indefinitely to follow the course of the ancients, as did the moderns before Descartes, and even for a little while afterwards, by adding some new curves to the small number of those which they had studied, the progress thus made, however rapid it might have been, would still be found, after a long series of ages, to be very inconsiderable in comparison with the general system of geometry, seeing the infinite variety of the forms which would still have remained to be studied. On the contrary, at each question resolved according to the method of the moderns, the number of geometrical problems to be resolved is then, once for all, diminished by so much with respect to all possible bodies. Another consideration is, that it resulted, from their complete want of general methods, that the ancient geometers, in all their investigations, were entirely abandoned to their own strength, without ever having the certainty of obtaining, sooner or later, any solution whatever. Though this imperfection of the science was eminently suited to call forth all their admirable sagacity, it necessarily rendered their progress extremely slow; we can form some idea of this by the considerable time which they employed in the study of the conic sections. Modern geometry, making the progress of our mind certain, permits us, on the contrary, to make the greatest possible use of the forces of our intelligence, which the ancients were often obliged to waste on very unimportant questions.

A no less important difference between the two systems appears when we come to consider geometry in the concrete point of view. Indeed, we have already remarked that the relation of the abstract to the concrete in geometry can be founded upon rational bases only so far as the investigations are made to bear directly upon all imaginable figures. In studying lines, only one by one, whatever may be the number, always necessarily very small, of those which we shall have considered, the application of such theories to figures really existing in nature will never have any other than an essentially accidental character, since there is nothing to assure us that these figures can really be brought under the abstract types considered by geometers.

Thus, for example, there is certainly something fortuitous in the happy relation established between the speculations of the Greek geometers upon the conic sections and the determination of the true planetary orbits. In continuing geometrical researches upon the same plan, there was no good reason for hoping for similar coincidences; and it would have been possible, in these special studies, that the researches of geometers should have been directed to abstract figures entirely incapable of any application, while they neglected others, susceptible perhaps of an important and immediate application. It is clear, at least, that nothing positively guaranteed the necessary applicability of geometrical speculations. It is quite another thing in the modern geometry. From the single circumstance that in it we proceed by general questions relating to any figures whatever, we have in advance the evident certainty that the figures really existing in the external world could in no case escape the appropriate theory if the geometrical phenomenon which it considers presents itself in them.

From these different considerations, we see that the ancient system of geometry wears essentially the character of the infancy of the science, which did not begin to become completely rational till after the philosophical resolution produced by Descartes. But it is evident, on the other hand, that geometry could not be at first conceived except in this *special* manner. *General* geometry would not have been possible, and its necessity could not even have been felt, if a long series of special labours on the most simple figures had not previously furnished bases for the conception of Descartes, and rendered apparent the impossibility of persisting indefinitely in the primitive geometrical philosophy.

The Ancient the Base of the Modern. From this last consideration we must infer that, although the geometry which I have called *general* must be now regarded as the only true dogmatical geometry, and that to which we shall chiefly confine ourselves, the other having no longer much more than an historical interest, nevertheless it is not possible to entirely dispense with *special* geometry in a rational exposition of the science. We undoubtedly need not borrow directly from ancient geometry all the results which it has furnished; but, from the very nature of the subject, it is necessarily impossible entirely to dispense with the ancient method, which will always serve as the preliminary basis of the science, dogmatically as well as historically. The reason of this is easy to understand. In fact, *general* geometry being essentially founded, as we shall soon establish, upon the employment of the calculus in the transformation of geometrical into analytical considerations, such a manner of proceeding could not take possession of the subject immediately at its origin. We know that the application of mathematical analysis, from its nature, can never commence any science whatever, since evidently it cannot be employed until the science has already been sufficiently cultivated to establish, with respect to the phenomena considered, some *equations* which can serve as starting points for the analytical operations. These fundamental equations being once discovered, analysis will enable us to deduce from them a multitude of consequences which it would have been previously impossible even to suspect; it will perfect the science to an immense degree, both with respect to the generality of its conceptions and to the complete co-ordination established between them. But mere mathematical analysis could never be sufficient to form the bases of any natural science, not even to demonstrate them anew when they have once been established. Nothing can dispense with the direct study of the subject, pursued up to the point of the discovery of precise relations.

We thus see that the geometry of the ancients will always have, by its nature, a primary part, absolutely necessary and more or less extensive, in the complete system of geometrical knowledge. It forms a rigorously indispensable introduction to *general* geometry. But it is to this that it must be limited in a completely dogmatic exposition. I will consider, then, directly, in the following chapter, this *special* or *preliminary* geometry restricted to exactly its necessary limits, in order to occupy myself thenceforth only with the philosophical examination of *general* or *definitive* geometry, the only one which is truly rational, and which at present essentially composes the science.

CHAPTER II.

ANCIENT OR SYNTHETIC GEOMETRY.

The geometrical method of the ancients necessarily constituting a preliminary department in the dogmatical system of geometry, designed to furnish *general* geometry with indispensable foundations, it is now proper to begin with determining wherein strictly consists this preliminary function of *special* geometry, thus reduced to the narrowest possible limits.

ITS PROPER EXTENT.

Lines; Polygons; Polyhedrons. In considering it under this point of view, it is easy to recognize that we might restrict it to the study of the right line alone for what concerns the geometry of *lines;* to the *quadrature* of rectilinear plane areas; and, lastly, to the *cubature* of bodies terminated by plane faces. The elementary propositions relating to these three fundamental questions form, in fact, the necessary starting point of all geometrical inquiries; they alone cannot be obtained except by a direct study of the subject; while, on the contrary, the complete theory of all other figures, even that of the circle, and of the surfaces and volumes which are connected with it, may at the present day be completely comprehended in the domain of *general* or *analytical* geometry; these primitive elements at once furnishing *equations* which are sufficient to allow of the application of the calculus to geometrical questions, which would not have been possible without this previous condition.

It results from this consideration that, in common practice, we give to *elementary* geometry more extent than would be rigorously necessary to it; since, besides the right line, polygons, and polyhedrons, we also include in it the circle and the "round" bodies; the study of which might, however, be as purely analytical as that, for example, of the conic sections. An unreflecting veneration for antiquity contributes to maintain this defect in method; but the best reason which can be given for it is the serious inconvenience for ordinary instruction which there would be in postponing, to so distant an epoch of mathematical education, the solution of several essential questions, which are susceptible of a direct and continual application to a great number of important uses. In fact, to proceed in the most rational manner, we should employ the integral calculus in obtaining the interesting results relating to the length or the area of the circle, or to the quadrature of the sphere, &c., which have been determined by the ancients from extremely simple considerations. This inconvenience would be of little importance with regard to the persons destined to study the whole of mathematical science, and the advantage of proceeding in a perfectly logical order would have a much greater comparative value. But the contrary case being the more frequent, theories so essential have necessarily been retained in elementary geometry. Perhaps the conic sections, the cycloid, &c., might be advantageously added in such cases.

Not to be farther restricted. While this preliminary portion of geometry, which cannot be founded on the application of the calculus, is reduced by its nature to a very limited series of fundamental researches, relating to the right line, polygonal areas, and polyhedrons, it is certain, on the other hand, that we cannot restrict it any more; although, by a veritable abuse of the spirit of analysis, it has been recently attempted to present the establishment of the principal theorems of elementary geometry under an algebraical point of view. Thus some have pretended to demonstrate, by simple abstract considerations of mathematical analysis, the constant relation which exists between the three angles of a rectilinear triangle, the fundamental proposition of the theory of similar triangles, that of parallelopipedons, &c.; in a word, precisely the only geometrical propositions which cannot be obtained except by a direct study of the subject, without the calculus being susceptible of having any part in it. Such aberrations are the unreflecting exaggerations of that natural and philosophical tendency which leads us to extend farther and farther the influence of analysis in mathematical studies. In mechanics, the pretended analytical demonstrations of the parallelogram of forces are of similar character.

The viciousness of such a manner of proceeding follows from the principles previously presented. We have already, in fact, recognized that, since the calculus is not, and cannot be, any thing but a means of deduction, it would indicate a radically false idea of it to wish to employ it in establishing the elementary foundations of any science whatever; for on what would the analytical reasonings in such an operation repose? A labour of this nature, very far from really perfecting the philosophical character of a science, would constitute a return towards the metaphysical age, in presenting real facts as mere logical abstractions.

When we examine in themselves these pretended analytical demonstrations of the fundamental propositions of elementary geometry, we easily verify their necessary want of meaning. They are all founded on a vicious manner of conceiving the principle of *homogeneity*, the true general idea of which was explained in the second chapter of the preceding book. These demonstrations suppose that this principle does not allow us to admit the coexistence in the same equation of numbers obtained by different concrete comparisons, which is evidently false, and contrary to the constant practice of geometers. Thus it is easy to recognize that, by employing the law of homogeneity in this arbitrary and illegitimate acceptation, we could succeed in "demonstrating," with quite as much apparent rigour, propositions whose absurdity is manifest at the first glance. In examining attentively, for example, the procedure by the aid of which it has been attempted to prove analytically that the sum of the three angles of any rectilinear triangle is constantly equal to two right angles, we see that it is founded on this preliminary principle that, if two triangles have two of their angles respectively equal, the third angle of the one will necessarily be equal to the third angle of the other. This first point being granted, the proposed relation is immediately deduced from it in a very exact and simple manner. Now the analytical consideration by which this previous proposition has been attempted to be established, is of such a nature that, if it could be correct, we could rigorously deduce from it, in reproducing it conversely, this palpable absurdity, that two sides of a triangle are sufficient, without any angle, for the entire determination of the third side. We may make analogous remarks on all the demonstrations of this sort, the sophisms of which will be thus verified in a perfectly apparent manner.

The more reason that we have here to consider geometry as being at the present day essentially analytical, the more necessary was it to guard against this abusive exaggeration of mathematical analysis, according to which all geometrical observation would be dispensed with, in establishing upon pure algebraical abstractions the very foundations of this natural science.

Attempted Demonstrations of Axioms, &c. Another indication that geometers have too much overlooked the character of a natural science which is necessarily inherent in geometry, appears from their vain attempts, so long made, to *demonstrate* rigorously, not by the aid of the calculus, but by means of certain constructions, several fundamental propositions of elementary geometry. Whatever may be effected, it will evidently be impossible to avoid sometimes recurring to simple and direct observation in geometry as a means of establishing various results: While, in this science, the phenomena which are considered are, by virtue of their extreme simplicity, much more closely connected with one another than those relating to any other physical science, some must still be found which cannot be deduced, and which, on the contrary, serve as starting points. It may be admitted that the greatest logical perfection of the science is to reduce these to the smallest number possible, but it would be absurd to pretend to make them completely disappear. I avow, moreover, that I find fewer real inconveniences in extending, a little beyond what would be strictly necessary, the number of these geometrical notions thus established by direct observation, provided they are sufficiently simple, than in making them the subjects of complicated and indirect demonstrations, even when these demonstrations may be logically irreproachable.

The true dogmatic destination of the geometry of the ancients, reduced to its least possible indispensable developments, having thus been characterized as exactly as possible, it is proper to consider summarily each of the principal parts of which it must be composed. I think that I may here limit myself to considering the first and the most extensive of these parts, that which has for its object the study of *the right line;* the two other sections, namely, the *quadrature of polygons* and the *cubature of polyhedrons*, from their limited extent, not being capable of giving rise to any philosophical consideration of any importance, distinct from those indicated in the preceding chapter with respect to the measure of areas and of volumes in general.

GEOMETRY OF THE RIGHT LINE.

The final question which we always have in view in the study of the right line, properly consists in determining, by means of one another, the different elements of any right-lined figure whatever; which enables us always to know indirectly the length and position of a right line, in whatever circumstances it may be placed. This fundamental problem is susceptible of two general solutions, the nature of which is quite distinct, the one *graphical*, the other *algebraic*. The first, though very imperfect, is that which must be first considered, because it is spontaneously derived from the direct study of the subject; the second, much more perfect in the most important respects, cannot be studied till afterwards, because it is founded upon the previous knowledge of the other.

GRAPHICAL SOLUTIONS.

The graphical solution consists in constructing at will the proposed figure, either with the same dimensions, or, more usually, with dimensions changed in any ratio whatever. The first mode need merely be mentioned as being the most simple and the one which would first occur to the mind, for it is evidently, by its nature, almost entirely incapable of application. The second is, on the contrary, susceptible of being most extensively and most usefully applied. We still make an important and continual use of it at the present day, not only to represent with exactness the forms of bodies and their relative positions, but even for the actual determination of geometrical magnitudes, when we do not need great precision. The ancients, in consequence of the imperfection of their geometrical knowledge, employed this procedure in a much more extensive manner, since it was for a long time the only one which they could apply, even in the most important precise determinations. It was thus, for example, that Aristarchus of Samos estimated the relative distance from the sun and from the moon to the earth, by making measurements on a triangle constructed as exactly as possible, so as to be similar to the right-angled triangle formed by the three bodies at the instant when the moon is in quadrature, and when an observation of the angle at the earth would consequently be sufficient to define the triangle. Archimedes himself, although he was the first to introduce calculated determinations into geometry, several times employed similar means. The formation of trigonometry did not cause this method to be entirely abandoned, although it greatly diminished its use; the Greeks and the Arabians continued to employ it for a great number of researches, in which we now regard the use of the calculus as indispensable.

This exact reproduction of any figure whatever on a different scale cannot present any great theoretical difficulty when all the parts of the proposed figure lie in the same plane. But if we suppose, as most frequently happens, that they are situated in different planes, we see, then, a new order of geometrical considerations arise. The artificial figure, which is constantly plane, not being capable, in that case, of being a perfectly faithful image of the real figure, it is necessary previously to fix with precision the mode of representation, which gives rise to different systems of *Projection.*

It then remains to be determined according to what laws the geometrical phenomena correspond in the two figures. This consideration generates a new series of geometrical investigations, the final object of which is properly to discover how we can replace constructions in relief by plane constructions. The ancients had to resolve several elementary questions of this kind for various cases in which we now employ spherical trigonometry, principally for different problems relating to the celestial sphere. Such was the object of their *analemmas*, and of the other plane figures which for a long time supplied the place of the calculus. We see by this that the ancients really knew the elements of what we now name *Descriptive Geometry*, although they did not conceive it in a distinct and general manner.

I think it proper briefly to indicate in this place the true philosophical character of this "Descriptive Geometry;" although, being essentially a science of application, it ought not to be included within the proper domain of this work.

DESCRIPTIVE GEOMETRY.

All questions of geometry of three dimensions necessarily give rise, when we consider their graphical solution, to a common difficulty which is peculiar to them; that of substituting for the different constructions in relief, which are necessary to resolve them directly, and which it is almost always impossible to execute, simple equivalent plane constructions, by means of which we finally obtain the same results. Without this indispensable transformation, every solution of this kind would be evidently incomplete and really inapplicable in practice, although theoretically the constructions in space are usually preferable as being more direct. It was in order to furnish general means for always effecting such a transformation that *Descriptive Geometry* was created, and formed into a distinct and homogeneous system, by the illustrious MONGE. He invented, in the first place, a uniform method of representing bodies by figures traced on a single plane, by the aid of *projections* on two different planes, usually perpendicular to each other, and one of which is supposed to turn about their common intersection so as to coincide with the other produced; in this system, or in any other equivalent to it, it is sufficient to regard points and lines as being determined by their projections, and surfaces by the projections of their generating lines. This being established, Monge—analyzing with profound sagacity the various partial labours of this kind which had before been executed by a number of incongruous procedures, and considering also, in a general and direct manner, in what any questions of that nature must consist—found that they could always be reduced to a very small number of invariable abstract problems, capable of being resolved separately, once for all, by uniform operations, relating essentially some to the contacts and others to the intersections of surfaces. Simple and entirely general methods for the graphical solution of these two orders of problems having been formed, all the geometrical questions which may arise in any of the various arts of construction—stone-cutting, carpentry, perspective, dialling, fortification, &c.—can henceforth be treated as simple particular cases of a single theory, the invariable application of which will always necessarily lead to an exact solution, which may be facilitated in practice by profiting by the peculiar circumstances of each case.

This important creation deserves in a remarkable degree to fix the attention of those philosophers who consider all that the human species has yet effected as a first step, and thus far the only really complete one, towards that general renovation of human labours, which must imprint upon all our arts a character of precision and of rationality, so necessary to their future progress. Such a revolution must, in fact, inevitably commence with that class of industrial labours, which is essentially connected with that science which is the most simple, the most perfect, and the most ancient. It cannot fail to extend hereafter, though with less facility, to all other practical operations. Indeed Monge himself, who conceived the true philosophy of the arts better than any one else, endeavoured to sketch out a corresponding system for the mechanical arts.

Essential as the conception of descriptive geometry really is, it is very important not to deceive ourselves with respect to its true destination, as did those who, in the excitement of its first discovery, saw in it a means of enlarging the general and abstract domain of rational geometry. The result has in no way answered to these mistaken hopes. And, indeed, is it not evident that descriptive geometry has no special value except as a science of application, and as forming the true special theory of the geometrical arts? Considered in its abstract relations, it could not introduce any truly distinct order of geometrical speculations. We must not forget that, in order that a geometrical question should fall within the peculiar domain of descriptive geometry, it must necessarily have been previously resolved by speculative geometry, the solutions of which then, as we have seen, always need to be prepared for practice in such a way as to supply the place of constructions in relief by plane constructions; a substitution which really constitutes the only characteristic function of descriptive geometry.

It is proper, however, to remark here, that, with regard to intellectual education, the study of descriptive geometry possesses an important philosophical peculiarity, quite independent of its high industrial utility. This is the advantage which it so pre-eminently offers—in habituating the mind to consider very complicated geometrical combinations in space, and to follow with precision their continual correspondence with the figures which are actually traced—of thus exercising to the utmost, in the most certain and precise manner, that important faculty of the human mind which is properly called "imagination," and which consists, in its elementary and positive acceptation, in representing to ourselves, clearly and easily, a vast and variable collection of ideal objects, as if they were really before us.

Finally, to complete the indication of the general nature of descriptive geometry by determining its logical character, we have to observe that, while it belongs to the geometry of the ancients by the character of its solutions, on the other hand it approaches the geometry of the moderns by the nature of the questions which compose it. These questions are in fact eminently remarkable for that generality which, as we saw in the preceding chapter, constitutes the true fundamental character of modern geometry; for the methods used are always conceived as applicable to any figures whatever, the peculiarity of each having only a purely secondary influence. The solutions of descriptive geometry are then graphical, like most of those of the ancients, and at the same time general, like those of the moderns.

After this important digression, we will pursue the philosophical examination of *special* geometry, always considered as reduced to its least possible development, as an indispensable introduction to *general* geometry. We have now sufficiently considered the *graphical* solution of the fundamental problem relating to the right line

—that is, the determination of the different elements of any right-lined figure by means of one another—and have now to examine in a special manner the *algebraic* solution.

ALGEBRAIC SOLUTIONS.

This kind of solution, the evident superiority of which need not here be dwelt upon, belongs necessarily, by the very nature of the question, to the system of the ancient geometry, although the logical method which is employed causes it to be generally, but very improperly, separated from it. We have thus an opportunity of verifying, in a very important respect, what was established generally in the preceding chapter, that it is not by the employment of the calculus that the modern geometry is essentially to be distinguished from the ancient. The ancients are in fact the true inventors of the present trigonometry, spherical as well as rectilinear; it being only much less perfect in their hands, on account of the extreme inferiority of their algebraical knowledge. It is, then, really in this chapter, and not, as it might at first be thought, in those which we shall afterwards devote to the philosophical examination of *general* geometry, that it is proper to consider the character of this important preliminary theory, which is usually, though improperly, included in what is called *analytical geometry*, but which is really only a complement of *elementary geometry* properly so called.

Since all right-lined figures can be decomposed into triangles, it is evidently sufficient to know how to determine the different elements of a triangle by means of one another, which reduces *polygonometry* to simple *trigonometry*.

TRIGONOMETRY.

The difficulty in resolving algebraically such a question as the above, consists essentially in forming, between the angles and the sides of a triangle, three distinct equations; which, when once obtained, will evidently reduce all trigonometrical problems to mere questions of analysis.

How to introduce Angles. In considering the establishment of these equations in the most general manner, we immediately meet with a fundamental distinction with respect to the manner of introducing the angles into the calculation, according as they are made to enter *directly*, by themselves or by the circular arcs which are proportional to them; or *indirectly*, by the chords of these arcs, which are hence called their *trigonometrical lines.* Of these two systems of trigonometry the second was of necessity the only one originally adopted, as being the only practicable one, since the condition of geometry made it easy enough to find exact relations between the sides of the triangles and the trigonometrical lines which represent the angles, while it would have been absolutely impossible at that epoch to establish equations between the sides and the angles themselves.

Advantages of introducing Trigonometrical Lines. At the present day, since the solution can be obtained by either system indifferently, that motive for preference no longer exists; but geometers have none the less persisted in following from choice the system primitively admitted from necessity; for, the same reason which enabled these trigonometrical equations to be obtained with much more facility, must, in like manner, as it is still more easy to conceive *à priori*, render these equations much more simple, since they then exist only between right lines, instead of being established between right lines and arcs of circles. Such a consideration has so much the more importance, as the question relates to formulas which are eminently elementary, and destined to be continually employed in all parts of mathematical science, as well as in all its various applications.

It may be objected, however, that when an angle is given, it is, in reality, always given by itself, and not by its trigonometrical lines; and that when it is unknown, it is its angular value which is properly to be determined, and not that of any of its trigonometrical lines. It seems, according to this, that such lines are only useless intermediaries between the sides and the angles, which have to be finally eliminated, and the introduction of which does not appear capable of simplifying the proposed research. It is indeed important to explain, with more generality and precision than is customary, the great real utility of this manner of proceeding.

Division of Trigonometry into two Parts. It consists in the fact that the introduction of these auxiliary magnitudes divides the entire question of trigonometry into two others essentially distinct, one of which has for its object to pass from the angles to their trigonometrical lines, or the converse, and the other of which proposes to determine the sides of the triangles by the trigonometrical lines of their angles, or the converse. Now the first of these two fundamental questions is evidently susceptible, by its nature, of being entirely treated and reduced to numerical tables once for all, in considering all possible angles, since it depends only upon those angles, and not at all upon the particular triangles in which they may enter in each case; while the solution of the second question must necessarily be renewed, at least in its arithmetical relations, for each new triangle which it is necessary to resolve. This is the reason why the first portion of the complete work, which would be precisely the most laborious, is no longer taken into the account, being always done in advance; while, if such a decomposition had not been performed, we would evidently have found ourselves under the obligation of recommencing the entire calculation in each particular case. Such is the essential property of the present trigonometrical system, which in fact would really present no actual advantage, if it was necessary to calculate continually the trigonometrical line of each angle to be considered, or the converse; the intermediate agency introduced would then be more troublesome than convenient.

In order to clearly comprehend the true nature of this conception, it will be useful to compare it with a still more important one, designed to produce an analogous effect either in its algebraic, or, still more, in its arithmetical relations—the admirable theory of *logarithms*. In examining in a philosophical manner the influence of this theory, we see in fact that its general result is to decompose all imaginable arithmetical operations into two distinct parts. The first and most complicated of these is capable of being executed in advance once for all (since it depends only upon the numbers to be considered, and not at all upon the infinitely different combinations into which they can enter), and consists in considering all numbers as assignable powers of a constant number. The second part of the calculation, which must of necessity be recommenced for each new formula which is to have its value determined, is thenceforth reduced to executing upon these exponents correlative operations which are infinitely more simple. I confine myself here to merely indicating this resemblance, which any one can carry out for himself.

We must besides observe, as a property (secondary at the present day, but all-important at its origin) of the trigonometrical system adopted, the very remarkable circumstance that the determination of angles by their trigonometrical lines, or the converse, admits of an arithmetical solution (the only one which is directly indispensable for the special destination of trigonometry) without the previous resolution of the corresponding algebraic question. It is doubtless to such a peculiarity that the ancients owed the possibility of knowing trigonometry. The investigation conceived in this way was so much the more easy, inasmuch as tables of chords (which the ancients naturally took as the trigonometrical lines) had been previously constructed for quite a different object, in the course of the labours of Archimedes on the rectification of the circle, from which resulted the actual determination of a certain series of chords; so that when Hipparchus subsequently invented trigonometry, he could confine himself to completing that operation by suitable intercalations; which shows clearly the connexion of ideas in that matter.

The Increase of such Trigonometrical Lines. To complete this philosophical sketch of trigonometry, it is proper now to observe that the extension of the same considerations which lead us to replace angles or arcs of circles by straight lines, with the view of simplifying our equations, must also lead us to employ concurrently several trigonometrical lines, instead of confining ourselves to one only (as did the ancients), so as to perfect this system by choosing that one which will be algebraically the most convenient on each occasion. In this point of view, it is clear that the number of these lines is in itself no ways limited; provided that they are determined by the arc, and that they determine it, whatever may be the law according to which they are derived from it, they are suitable to be substituted for it in the equations. The Arabians, and subsequently the moderns, in confining themselves to the most simple constructions, have carried to four or five the number of *direct* trigonometrical lines, which might be extended much farther.

But instead of recurring to geometrical formations, which would finally become very complicated, we conceive with the utmost facility as many new trigonometrical lines as the analytical transformations may require, by means of a remarkable artifice, which is not usually apprehended in a sufficiently general manner. It consists in not directly multiplying the trigonometrical lines appropriate to each arc considered, but in introducing new ones, by considering this arc as indirectly determined by all lines relating to an arc which is a very simple function of the first. It is thus, for example, that, in order to calculate an angle with more facility, we will determine, instead of its sine, the sine of its half, or of its double, &c. Such a creation of *indirect* trigonometrical lines is evidently much more fruitful than all the direct geometrical methods for obtaining new ones. We may accordingly say that the number of trigonometrical lines actually employed at the present day by geometers is in reality unlimited, since at every instant, so to say, the transformations of analysis may lead us to augment it by the method which I have just indicated. Special names, however, have been given to those only of these *indirect* lines which refer to the complement of the primitive arc, the others not occurring sufficiently often to render such denominations necessary; a circumstance which has caused a common misconception of the true extent of the system of trigonometry.

Study of their Mutual Relations. This multiplicity of trigonometrical lines evidently gives rise to a third fundamental question in trigonometry, the study of the relations which exist between these different lines; since, without such a knowledge, we could not make use, for our analytical necessities, of this variety of auxiliary magnitudes, which, however, have no other destination. It is clear, besides, from the consideration just indicated, that this essential part of trigonometry, although simply preparatory, is, by its nature, susceptible of an indefinite extension when we view it in its entire generality, while the two others are circumscribed within rigorously defined limits.

It is needless to add that these three principal parts of trigonometry have to be studied in precisely the inverse order from that in which we have seen them necessarily derived from the general nature of the subject; for the third is evidently independent of the two others, and the second, of that which was first presented—the resolution of triangles, properly so called—which must for that reason be treated in the last place; which rendered so much the more important the consideration of their natural succession and logical relations to one another.

It is useless to consider here separately *spherical trigonometry*, which cannot give rise to any special philosophical consideration; since, essential as it is by the importance and the multiplicity of its uses, it can be treated at the present day only as a simple application of rectilinear trigonometry, which furnishes directly its fundamental equations, by substituting for the spherical triangle the corresponding trihedral angle.

This summary exposition of the philosophy of trigonometry has been here given in order to render apparent, by an important example, that rigorous dependence and those successive ramifications which are presented by what are apparently the most simple questions of elementary geometry.

Having thus examined the peculiar character of *special* geometry reduced to its only dogmatic destination, that of furnishing to general geometry an indispensable preliminary basis, we have now to give all our attention to the true science of geometry, considered as a whole, in the most rational manner. For that purpose, it is necessary to carefully examine the great original idea of Descartes, upon which it is entirely founded. This will be the object of the following chapter.

CHAPTER III.

MODERN OR ANALYTICAL GEOMETRY.

General (or *Analytical*) geometry being entirely founded upon the transformation of geometrical considerations into equivalent analytical considerations, we must begin with examining directly and in a thorough manner the beautiful conception by which Descartes has established in a uniform manner the constant possibility of such a co-relation. Besides its own extreme importance as a means of highly perfecting geometrical science, or, rather, of establishing the whole of it on rational bases, the philosophical study of this admirable conception must have so much the greater interest in our eyes from its characterizing with perfect clearness the general method to be employed in organizing the relations of the abstract to the concrete in mathematics, by the analytical representation of natural phenomena. There is no conception, in the whole philosophy of mathematics which better deserves to fix all our attention.

ANALYTICAL REPRESENTATION OF FIGURES.

In order to succeed in expressing all imaginable geometrical phenomena by simple analytical relations, we must evidently, in the first place, establish a general method for representing analytically the subjects themselves in which these phenomena are found, that is, the lines or the surfaces to be considered. The *subject* being thus habitually considered in a purely analytical point of view, we see how it is thenceforth possible to conceive in the same manner the various *accidents* of which it is susceptible.

In order to organize the representation of geometrical figures by analytical equations, we must previously surmount a fundamental difficulty; that of reducing the general elements of the various conceptions of geometry to simply numerical ideas; in a word, that of substituting in geometry pure considerations of *quantity* for all considerations of *quality*.

Reduction of Figure to Position. For this purpose let us observe, in the first place, that all geometrical ideas relate necessarily to these three universal categories: the *magnitude*, the *figure*, and the *position* of the extensions to be considered. As to the first, there is evidently no difficulty; it enters at once into the ideas of numbers. With relation to the second, it must be remarked that it will always admit of being reduced to the third. For the figure of a body evidently results from the mutual position of the different points of which it is composed, so that the idea of position necessarily comprehends that of figure, and every circumstance of figure can be translated by a circumstance of position. It is in this way, in fact, that the human mind has proceeded in order to arrive at the analytical representation of geometrical figures, their conception relating directly only to positions. All the elementary difficulty is then properly reduced to that of referring ideas of situation to ideas of magnitude. Such is the direct destination of the preliminary conception upon which Descartes has established the general system of analytical geometry.

His philosophical labour, in this relation, has consisted simply in the entire generalization of an elementary operation, which we may regard as natural to the human mind, since it is performed spontaneously, so to say, in all minds, even the most uncultivated. Thus, when we have to indicate the situation of an object without directly pointing it out, the method which we always adopt, and evidently the only one which can be employed, consists in referring that object to others which are known, by assigning the magnitude of the various geometrical elements, by which we conceive it connected with the known objects. These elements constitute what Descartes, and after him all geometers, have called the *co-ordinates* of each point considered. They are necessarily two in number, if it is known in advance in what plane the point is situated; and three, if it may be found indifferently in any region of space. As many different constructions as can be imagined for determining the position of a point, whether on a plane or in space, so many distinct systems of co-ordinates may be conceived; they are consequently susceptible of being multiplied to infinity. But, whatever may be the system adopted, we shall always have reduced the ideas of situation to simple ideas of magnitude, so that we will consider the change in the position of a point as produced by mere numerical variations in the values of its co-ordinates.

Determination of the Position of a Point. Considering at first only the least complicated case, that of *plane geometry*, it is in this way that we usually determine the position of a point on a plane, by its distances from two fixed right lines considered as known, which are called *axes*, and which are commonly supposed to be perpendicular to each other. This system is that most frequently adopted, because of its simplicity; but geometers employ occasionally an infinity of others. Thus the position of a point on a plane may be determined, 1°, by its distances from two fixed points; or, 2°, by its distance from a single fixed point, and the direction of that distance, estimated by the greater or less angle which it makes with a fixed right line, which constitutes the system of what are called *polar* co-ordinates, the most frequently used after the system first mentioned; or, 3°, by the angles which the right lines drawn from the variable point to two fixed points make with the right line which joins these last; or, 4°, by the distances from that point to a fixed right line and a fixed point, &c. In a word, there is no geometrical figure whatever from which it is not possible to deduce a certain system of co-ordinates more or less susceptible of being employed.

A general observation, which it is important to make in this connexion, is, that every system of co-ordinates is equivalent to determining a point, in plane geometry, by the intersection of two lines, each of which is subjected to certain fixed conditions of determination; a single one of these conditions remaining variable, sometimes the one, sometimes the other, according to the system considered. We could not, indeed, conceive any other means of constructing a point than to mark it by the meeting of two lines. Thus, in the most common system, that of *rectilinear co-ordinates,* properly so called, the point is determined by the intersection of two right lines, each of which remains constantly parallel to a fixed axis, at a greater or less distance from it; in the *polar* system, the position of the point is marked by the meeting of a circle, of variable radius and fixed centre, with a movable right line compelled to turn about this centre: in other systems, the required point might be designated by the intersection of two circles, or of any other two lines, &c. In a word, to assign the value of one of the co-ordinates of a point in any system whatever, is always necessarily equivalent to determining a certain line on which that point must be situated. The geometers of antiquity had already made this essential remark, which served as the base of their method of geometrical *loci*, of which they made so happy a use to direct their researches in the resolution of *determinate* problems, in considering separately the influence of each of the two conditions by which was defined each point constituting the object, direct or indirect, of the proposed question. It was the general systematization of this method which was the immediate motive of the labours of Descartes, which led him to create analytical geometry.

After having clearly established this preliminary conception—by means of which ideas of position, and thence, implicitly, all elementary geometrical conceptions are capable of being reduced to simple numerical considerations—it is easy to form a direct conception, in its entire generality, of the great original idea of Descartes, relative to the analytical representation of geometrical figures: it is this which forms the special object of this chapter. I will continue to consider at first, for more facility, only geometry of two dimensions, which alone was treated by Descartes; and will afterwards examine separately, under the same point of view, the theory of surfaces and curves of double curvature.

PLANE CURVES.

Expression of Lines by Equations. In accordance with the manner of expressing analytically the position of a point on a plane, it can be easily established that, by whatever property any line may be defined, that definition always admits of being replaced by a corresponding equation between the two variable co-ordinates of the point which describes this line; an equation which will be thenceforth the analytical representation of the proposed line, every phenomenon of which will be translated by a certain algebraic modification of its equation. Thus, if we suppose that a point moves on a plane without its course being in any manner determined, we shall evidently have to regard its co-ordinates, to whatever system they may belong, as two variables entirely independent of one another. But if, on the contrary, this point is compelled to describe a certain line, we shall necessarily be compelled to conceive that its co-ordinates, in all the positions which it can take, retain a certain permanent and precise relation to each other, which is consequently susceptible of being expressed by a suitable equation; which will become the very clear and very rigorous analytical definition of the line under consideration, since it will express an algebraical property belonging exclusively to the co-ordinates of all the points of this line. It is clear, indeed, that when a point is not subjected to any condition, its situation is not determined except in giving at once its two co-ordinates, independently of each other; while, when the point must continue upon a defined line, a single co-ordinate is sufficient for completely fixing its position. The second co-ordinate is then a determinate *function* of the first; or, in other words, there must exist between them a certain *equation*, of a nature corresponding to that of the line on which the point is compelled to remain. In a word, each of the co-ordinates of a point requiring it to be situated on a certain line, we conceive reciprocally that the condition, on the part of a point, of having to belong to a line defined in any manner whatever, is equivalent to assigning the value of one of the two co-ordinates; which is found in that case to be entirely dependent on the other. The analytical relation which expresses this dependence may be more or less difficult to discover, but it must evidently be always conceived to exist, even in the cases in which our present means may be insufficient to make it known. It is by this simple consideration that we may demonstrate, in an entirely general manner—independently of the particular verifications on which this fundamental conception is ordinarily established for each special definition of a line—the necessity of the analytical representation of lines by equations.

Expression of Equations by Lines. Taking up again the same reflections in the inverse direction, we could show as easily the geometrical necessity of the representation of every equation of two variables, in a determinate system of co-ordinates, by a certain line; of which such a relation would be, in the absence of any other known property, a very characteristic definition, the scientific destination of which will be to fix the attention directly upon the general course of the solutions of the equation, which will thus be noted in the most striking and the most simple manner. This picturing of equations is one of the most important fundamental advantages of analytical geometry, which has thereby reacted in the highest degree upon the general perfecting of analysis itself; not only by assigning to purely abstract researches a clearly determined object and an inexhaustible career, but, in a still more direct relation, by furnishing a new philosophical medium for analytical meditation which could not be replaced by any other. In fact, the purely algebraic discussion of an equation undoubtedly makes known its solutions in the most precise manner, but in considering them only one by one, so that in this way no general view of them could be obtained, except as the final result of a long and laborious series of numerical comparisons. On the other hand, the geometrical *locus* of the equation, being only designed to represent distinctly and with perfect clearness the summing up of all these comparisons, permits it to be directly considered, without paying any attention to the details which have furnished it. It can thereby suggest to our mind general analytical views, which we should have arrived at with much difficulty in any other manner, for want of a means of clearly characterizing their object. It is evident, for example, that the simple inspection of the logarithmic curve, or of the curve $y = \sin. x$, makes us perceive much more distinctly the general manner of the variations of logarithms with respect to their numbers, or of sines with respect to their arcs, than could the most attentive study of a table of logarithms or of natural sines. It is well known that this method has become entirely elementary at the present day, and that it is employed whenever it is desired to get a clear idea of the general character of the law which reigns in a series of precise observations of any kind whatever.

Any Change in the Line causes a Change in the Equation. Returning to the representation of lines by equations, which is our principal object, we see that this representation is, by its nature, so faithful, that the line could not experience any modification, however slight it might be, without causing a corresponding change in the equation. This perfect exactitude even gives rise oftentimes to special difficulties; for since, in our system of analytical geometry, the mere displacements of lines affect the equations, as well as their real variations in magnitude or form, we should be liable to confound them with one another in our analytical expressions, if geometers had not discovered an ingenious method designed expressly to always distinguish them. This method is founded on this principle, that although it is impossible to change analytically at will the position of a line with respect to the axes of the co-ordinates, we can change in any manner whatever the situation of the axes themselves, which evidently amounts to the same; then, by the aid of the very simple general formula by which this transformation of the axes is produced, it becomes easy to discover whether two different equations are the analytical expressions of only the same line differently situated, or refer to truly distinct geometrical loci; since, in the former case, one of them will pass into the other by suitably changing the axes or the other constants of the system of co-ordinates employed. It must, moreover, be remarked on this subject, that general inconveniences of this nature seem to be absolutely inevitable in analytical geometry; for, since the ideas of position are, as we have seen, the only geometrical ideas immediately reducible to numerical considerations, and the conceptions of figure cannot be thus reduced, except by seeing in them relations of situation, it is impossible for analysis to escape confounding, at first, the phenomena of figure with simple phenomena of position, which alone are directly expressed by the equations.

Every Definition of a Line is an Equation. In order to complete the philosophical explanation of the fundamental conception which serves as the base of analytical geometry, I think that I should here indicate a new general consideration, which seems to me particularly well adapted for putting in the clearest point of view this necessary representation of lines by equations with two variables. It consists in this, that not only, as we have shown, must every defined line necessarily give rise to a certain equation between the two co-ordinates of any one of its points, but, still farther, every definition of a line may be regarded as being already of itself an equation of that line in a suitable system of co-ordinates.

It is easy to establish this principle, first making a preliminary logical distinction with respect to different kinds of definitions. The rigorously indispensable condition of every definition is that of distinguishing the object defined from all others, by assigning to it a property which belongs to it exclusively. But this end may be generally attained in two very different ways; either by a definition which is simply *characteristic*, that is, indicative of a property which, although truly exclusive, does not make known the mode of generation of the object; or by a definition which is really *explanatory*, that is, which characterizes the object by a property which expresses one of its modes of generation. For example, in considering the circle as the line, which, under the same contour, contains the greatest area, we have evidently a definition of the first kind; while in choosing the property of its having all its points eqally distant from a fixed point, we have a definition of the second kind. It is, besides, evident, as a general principle, that even when any object whatever is known at first only by a *characteristic* definition, we ought, nevertheless, to regard it as susceptible of *explanatory* definitions, which the farther study of the object would necessarily lead us to discover.

This being premised, it is clear that the general observation above made, which represents every definition of a line as being necessarily an equation of that line in a certain system of co-ordinates, cannot apply to definitions which are simply *characteristic;* it is to be understood only of definitions which are truly *explanatory*. But, in considering only this class, the principle is easy to prove. In fact, it is evidently impossible to define the generation of a line without specifying a certain relation between the two simple motions of translation or of rotation, into which the motion of the point which describes it will be decomposed at each instant. Now if we form the most general conception of what constitutes *a system of co-ordinates*, and admit all possible systems, it is clear that such a relation will be nothing else but the *equation* of the proposed line, in a system of co-ordinates of a nature corresponding to that of the mode of generation considered. Thus, for example, the common definition of the *circle* may evidently be regarded as being immediately the *polar equation* of this curve, taking the centre of the circle for the pole. In the same way, the elementary definition of the *ellipse* or of the *hyperbola*—as being the curve generated by a point which moves in such a manner that the sum or the difference of its distances from two fixed points remains constant—gives at once, for either the one or the other curve, the equation $y+x=c$, taking for the system of co-ordinates that in which the position of a point would be determined by its distances from two fixed points, and choosing for these poles the two given foci. In like manner, the common definition of any *cycloid* would furnish directly, for that curve, the equation $y=mx$; adopting as the co-ordinates of each point the arc which it marks upon a circle of invariable radius, measuring from the point of contact of that circle with a fixed line, and the rectilinear distance from that point of contact to a certain origin taken on that right line. We can make analogous and equally easy verifications with respect to the customary definitions of spirals, of epicycloids, &c. We shall constantly find that there exists a certain system of co-ordinates, in which we immediately obtain a very simple equation of the proposed line, by merely writing algebraically the condition imposed by the mode of generation considered.

Besides its direct importance as a means of rendering perfectly apparent the necessary representation of every line by an equation, the preceding consideration seems to me to possess a true scientific utility, in characterizing with precision the principal general difficulty which occurs in the actual establishment of these equations, and in consequently furnishing an interesting indication with respect to the course to be pursued in inquiries of this kind, which, by their nature, could not admit of complete and invariable rules. In fact, since any definition whatever of a line, at least among those which indicate a mode of generation, furnishes directly the equation of that line in a certain system of co-ordinates, or, rather, of itself constitutes that equation, it follows that the difficulty which we often experience in discovering the equation of a curve, by means of certain of its characteristic properties, a difficulty which is sometimes very great, must proceed essentially only from the commonly imposed condition of expressing this curve analytically by the aid of a designated system of co-ordinates, instead of admitting indifferently all possible systems. These different systems cannot be regarded in analytical geometry as being all equally suitable; for various reasons, the most important of which will be hereafter discussed, geometers think that curves should almost always be referred, as far as is possible, to *rectilinear co-ordinates*, properly so called. Now we see, from what precedes, that in many cases these particular co-ordinates will not be those with reference to which the equation of the curve will be found to be directly established by the proposed definition. The principal difficulty presented by the formation of the equation of a line really consists, then, in general, in a certain transformation of co-ordinates. It is undoubtedly true that this consideration does not subject the establishment of these equations to a truly complete general method, the success of which is always certain; which, from the very nature of the subject, is evidently chimerical: but such a view may throw much useful light upon the course which it is proper to adopt, in order to arrive at the end proposed. Thus, after having in the first place formed the preparatory equation, which is spontaneously derived from the definition which we are considering, it will be necessary, in order to obtain the equation belonging to the system of co-ordinates which must be finally admitted, to endeavour to express in a function of these last co-ordinates those which naturally correspond to the given mode of generation. It is upon this last labour that it is evidently impossible to give invariable and precise precepts. We can only say that we shall have so many more resources in this matter as we shall know more of true analytical geometry, that is, as we shall know the algebraical expression of a greater number of different algebraical phenomena.

CHOICE OF CO-ORDINATES.

In order to complete the philosophical exposition of the conception which serves as the base of analytical geometry, I have yet to notice the considerations relating to the choice of the system of co-ordinates which is in general the most suitable. They will give the rational explanation of the preference unanimously accorded to the ordinary rectilinear system ; a preference which has hitherto been rather the effect of an empirical sentiment of the superiority of this system, than the exact result of a direct and thorough analysis.

Two different Points of View. In order to decide clearly between all the different systems of co-ordinates, it is indispensable to distinguish with care the two general points of view, the converse of one another, which belong to analytical geometry ; namely, the relation of algebra to geometry, founded upon the representation of lines by equations ; and, reciprocally, the relation of geometry to algebra, founded on the representation of equations by lines.

It is evident that in every investigation of general geometry these two fundamental points of view are of necessity always found combined, since we have always to pass alternately, and at insensible intervals, so to say, from geometrical to analytical considerations, and from analytical to geometrical considerations. But the necessity of here temporarily separating them is none the less real; for the answer to the question of method which we are examining is, in fact, as we shall see presently, very far from being the same in both these relations, so that without this distinction we could not form any clear idea of it.

1. *Representation of Lines by Equations.* Under *the first point of view*—the representation of lines by equations—the only reason which could lead us to prefer one system of co-ordinates to another would be the greater simplicity of the equation of each line, and greater facility in arriving at it. Now it is easy to see that there does not exist, and could not be expected to exist, any system of co-ordinates deserving in that respect a constant preference over all others. In fact, we have above remarked that for each geometrical definition proposed we can conceive a system of co-ordinates in which the equation of the line is obtained at once, and is necessarily found to be also very simple; and this system, moreover, inevitably varies with the nature of the characteristic property under consideration. The rectilinear system could not, therefore, be constantly the most advantageous for this object, although it may often be very favourable; there is probably no system which, in certain particular cases, should not be preferred to it, as well as to every other.

2. *Representation of Equations by Lines.* It is by no means so, however, under the *second point of view.* We can, indeed, easily establish, as a general principle, that

the ordinary rectilinear system must necessarily be better adapted than any other to the representation of equations by the corresponding geometrical *loci;* that is to say, that this representation is constantly more simple and more faithful in it than in any other.

Let us consider, for this object, that, since every system of co-ordinates consists in determining a point by the intersection of two lines, the system adapted to furnish the most suitable geometrical *loci* must be that in which these two lines are the simplest possible; a consideration which confines our choice to the *rectilinear* system. In truth, there is evidently an infinite number of systems which deserve that name, that is to say, which employ only right lines to determine points, besides the ordinary system which assigns the distances from two fixed lines as co-ordinates; such, for example, would be that in which the co-ordinates of each point should be the two angles which the right lines, which go from that point to two fixed points, make with the right line, which joins these last points: so that this first consideration is not rigorously sufficient to explain the preference unanimously given to the common system. But in examining in a more thorough manner the nature of every system of co-ordinates, we also perceive that each of the two lines, whose meeting determines the point considered, must necessarily offer at every instant, among its different conditions of determination, a single variable condition, which gives rise to the corresponding co-ordinate, all the rest being fixed, and constituting the *axes* of the system, taking this term in its most extended mathematical acceptation. The variation is indispensable, in order that we may be able to consider all possible positions; and the fixity is no less so, in order that there may exist means of comparison. Thus, in all *rectilinear* systems, each of the two right lines will be subjected to a fixed condition, and the ordinate will result from the variable condition.

Superiority of rectilinear Co-ordinates. From these considerations it is evident, as a general principle, that the most favourable system for the construction of geometrical *loci* will necessarily be that in which the variable condition of each right line shall be the simplest possible; the fixed condition being left free to be made complex, if necessary to attain that object. Now, of all possible manners of determining two movable right lines, the easiest to follow geometrically is certainly that in which, the direction of each right line remaining invariable, it only approaches or recedes, more or less, to or from a constant axis. It would be, for example, evidently more difficult to figure to one's self clearly the changes of place of a point which is determined by the intersection of two right lines, which each turn around a fixed point, making a greater or smaller angle with a certain axis, as in the system of co-ordinates previously noticed. Such is the true general explanation of the fundamental property possessed by the common rectilinear system, of being better adapted than any other to the geometrical representation of equations, inasmuch as it is that one in which it is the easiest to conceive the change of place of a point resulting from the change in the value of its co-ordinates. In order to feel clearly all the force of this consideration, it would be sufficient to carefully compare this system with the polar system, in which this geometrical image, so simple and so easy to follow, of two right lines moving parallel, each one of them, to its corresponding axis, is replaced by the complicated picture of an infinite series of concentric circles, cut by a right line compelled to turn about a fixed point. It is, moreover, easy to conceive in advance what must be the extreme importance to analytical geometry of a property so profoundly elementary, which, for that reason, must be recurring at every instant, and take a progressively increasing value in all labours of this kind.

Perpendicularity of the Axes. In pursuing farther the consideration which demonstrates the superiority of the ordinary system of co-ordinates over any other as to the representation of equations, we may also take notice of the utility for this object of the common usage of taking the two axes perpendicular to each other, whenever possible, rather than with any other inclination. As regards the representation of lines by equations, this secondary circumstance is no more universally proper than we have seen the general nature of the system to be; since, according to the particular occasion, any other inclination of the axes may deserve our preference in that respect. But, in the inverse point of view, it is easy to see that rectangular axes constantly permit us to represent equations in a more simple and even more faithful manner; for, with oblique axes, space being divided by them into regions which no longer have a perfect identity, it follows that, if the geometrical *locus* of the equation extends into all these regions at once, there will be presented, by reason merely of this inequality of the angles, differences of figure which do not correspond to any analytical diversity, and will necessarily alter the rigorous exactness of the representation, by being confounded with the proper results of the algebraic comparisons. For example, an equation like $x^m + y^m = c$, which, by its perfect symmetry, should evidently give a curve composed of four identical quarters, will be represented, on the contrary, if we take axes not rectangular, by a geometric *locus*, the four parts of which will be unequal. It is plain that the only means of avoiding all inconveniences of this kind is to suppose the angle of the two axes to be a right angle.

The preceding discussion clearly shows that, although the ordinary system of rectilinear co-ordinates has no constant superiority over all others in one of the two fundamental points of view which are continually combined in analytical geometry, yet as, on the other hand, it is not constantly inferior, its necessary and absolute greater aptitude for the representation of equations must cause it to generally receive the preference; although it may evidently happen, in some particular cases, that the necessity of simplifying equations and of obtaining them more easily may determine geometers to adopt a less perfect system. The rectilinear system is, therefore, the one by means of which are ordinarily constructed the most essential theories of general geometry, intended to express analytically the most important geometrical phenomena. When it is thought necessary to choose some other, the polar system is almost always the one which is fixed upon, this system being of a nature sufficiently opposite to that of the rectilinear system to cause the equations, which are too complicated with respect to the latter, to become, in general, sufficiently simple with respect to the other. Polar co-ordinates, moreover, have often the advantage of admitting of a more direct and natural concrete signification; as is the case in mechanics, for the geometrical questions to which the theory of circular movement gives rise, and in almost all the cases of celestial geometry.

In order to simplify the exposition, we have thus far considered the fundamental conception of analytical geometry only with respect to *plane curves*, the general study of which was the only object of the great philosophical renovation produced by Descartes. To complete this important explanation, we have now to show summarily how this elementary idea was extended by Clairaut, about a century afterwards, to the general study of *surfaces* and *curves of double curvature*. The considerations which have been already given will permit me to limit myself on this subject to the rapid examination of what is strictly peculiar to this new case.

SURFACES.

Determination of a Point in Space. The complete analytical determination of a point in space evidently requires the values of three co-ordinates to be assigned; as, for example, in the system which is generally adopted, and which corresponds to the *rectilinear* system of plane geometry, distances from the point to three fixed planes, usually perpendicular to one another; which presents the point as the intersection of three planes whose direction is invariable. We might also employ the distances from the movable point to three fixed points, which would determine it by the intersection of three spheres with a common centre. In like manner, the position of a point would be defined by giving its distance from a fixed point, and the direction of that distance, by means of the two angles which this right line makes with two invariable axes; this is the *polar* system of geometry of three dimensions; the point is then constructed by the intersection of a sphere having a fixed centre, with two right cones with circular bases, whose axes and common summit do not change. In a word, there is evidently, in this case at least, the same infinite variety among the various possible systems of co-ordinates which we have already observed in geometry of two dimensions. In general, we have to conceive a point as being always determined by the intersection of any three surfaces whatever, as it was in the former case by that of two lines: each of these three surfaces has, in like manner, all its conditions of determination constant, excepting one, which gives rise to the corresponding co-ordinates, whose peculiar geometrical influence is thus to constrain the point to be situated upon that surface.

This being premised, it is clear that if the three coordinates of a point are entirely independent of one another, that point can take successively all possible positions in space. But if the point is compelled to remain upon a certain surface defined in any manner whatever, then two co-ordinates are evidently sufficient for determining its situation at each instant, since the proposed surface will take the place of the condition imposed by the third co-ordinate. We must then, in this case, under the analytical point of view, necessarily conceive this last co-ordinate as a determinate function of the two others, these latter remaining perfectly independent of each other. Thus there will be a certain equation between the three variable co-ordinates, which will be permanent, and which will be the only one, in order to correspond to the precise degree of indetermination in the position of the point.

Expression of Surfaces by Equations. This equation, more or less easy to be discovered, but always possible, will be the analytical definition of the proposed surface, since it must be verified for all the points of that surface, and for them alone. If the surface undergoes any change whatever, even a simple change of place, the equation must undergo a more or less serious corresponding modification. In a word, all geometrical phenomena relating to surfaces will admit of being translated by certain equivalent analytical conditions appropriate to equations of three variables; and in the establishment and interpretation of this general and necessary harmony will essentially consist the science of analytical geometry of three dimensions.

Expression of Equations by Surfaces. Considering next this fundamental conception in the inverse point of view, we see in the same manner that every equation of three variables may, in general, be represented geometrically by a determinate surface, primitively defined by the very characteristic property, that the co-ordinates of all its points always retain the mutual relation enunciated in this equation. This geometrical locus will evidently change, for the same equation, according to the system of co-ordinates which may serve for the construction of this representation. In adopting, for example, the rectilinear system, it is clear that in the equation between the three variables, x, y, z, every particular value attributed to z will give an equation between x and y, the geometrical locus of which will be a certain line situated in a plane parallel to the plane of x and y, and at a distance from this last equal to the value of z; so that the complete geometrical locus will present itself as composed of an infinite series of lines superimposed in a series of parallel planes (excepting the interruptions which may exist), and will consequently form a veritable surface. It would be the same in considering any other system of co-ordinates, although the geometrical construction of the equation becomes more difficult to follow.

Such is the elementary conception, the complement of the original idea of Descartes, on which is founded general geometry relative to surfaces. It would be useless to take up here directly the other considerations which have been above indicated, with respect to lines, and which any one can easily extend to surfaces; whether to show that every definition of a surface by any method of generation whatever is really a direct equation of that surface in a certain system of co-ordinates, or to determine among all the different systems of possible co-ordinates that one which is generally the most convenient. I will only add, on this last point, that the necessary superiority of the ordinary rectilinear system, as to the representation of equations, is evidently still more marked in analytical geometry of three dimensions than in that of two, because of the incomparably greater geometrical complication which would result from the choice of any other system. This can be verified in the most striking manner by considering the polar system in particular, which is the most employed after the ordinary rectilinear system, for surfaces as well as for plane curves, and for the same reasons.

In order to complete the general exposition of the fundamental conception relative to the analytical study of surfaces, a philosophical examination should be made of a final improvement of the highest importance, which Monge has introduced into the very elements of this theory, for the classification of surfaces in natural families, established according to the mode of generation, and expressed algebraically by common differential equations, or by finite equations containing arbitrary functions.

CURVES OF DOUBLE CURVATURE.

Let us now consider the last elementary point of view of analytical geometry of three dimensions ; that relating to the algebraic representation of curves considered in space, in the most general manner. In continuing to follow the principle which has been constantly employed, that of the degree of indetermination of the geometrical locus, corresponding to the degree of independence of the variables, it is evident, as a general principle, that when a point is required to be situated upon some certain curve, a single co-ordinate is enough for completely determining its position, by the intersection of this curve with the surface which results from this co-ordinate. Thus, in this case, the two other co-ordinates of the point must be conceived as functions necessarily determinate and distinct from the first. It follows that every line, considered in space, is then represented analytically, no longer by a single equation, but by the system of two equations between the three co-ordinates of any one of its points. It is clear, indeed, from another point of view, that since each of these equations, considered separately, expresses a certain surface, their combination presents the proposed line as the intersection of two determinate surfaces.

Such is the most general manner of conceiving the algebraic representation of a line in analytical geometry of three dimensions. This conception is commonly considered in too restricted a manner, when we confine ourselves to considering a line as determined by the system of its two *projections* upon two of the co-ordinate planes; a system characterized, analytically, by this peculiarity, that each of the two equations of the line then contains only two of the three co-ordinates, instead of simultaneously including the three variables. This consideration, which consists in regarding the line as the intersection of two cylindrical surfaces parallel to two of the three axes of the co-ordinates, besides the inconvenience of being confined to the ordinary rectilinear system, has the fault, if we strictly confine ourselves to it, of introducing useless difficulties into the analytical representation of lines, since the combination of these two cylinders would evidently not be always the most suitable for forming the equations of a line. Thus, considering this fundamental notion in its entire generality, it will be necessary in each case to choose, from among the infinite number of couples of surfaces, the intersection of which might produce the proposed curve, that one which will lend itself the best to the establishment of equations, as being composed of the best known surfaces. Thus, if the problem is to express analytically a circle in space, it will evidently be preferable to consider it as the intersection of a sphere and a plane, rather than as proceeding from any other combination of surfaces which could equally produce it.

In truth, this manner of conceiving the representation of lines by equations, in analytical geometry of three dimensions, produces, by its nature, a necessary inconvenience, that of a certain analytical confusion, consisting in this: that the same line may thus be expressed, with the same system of co-ordinates, by an infinite number of different couples of equations, on account of the infinite number of couples of surfaces which can form it; a circumstance which may cause some difficulties in recognizing this line under all the algebraical disguises of which it admits. But there exists a very simple method for causing this inconvenience to disappear; it consists in giving up the facilities which result from this variety of geometrical constructions. It suffices, in fact, whatever may be the analytical system primitively established for a certain line, to be able to deduce from it the system corresponding to a single couple of surfaces uniformly generated; as, for example, to that of the two cylindrical surfaces which *project* the proposed line upon two of the co-ordinate planes; surfaces which will evidently be always identical, in whatever manner the line may have been obtained, and which will not vary except when that line itself shall change. Now, in choosing this fixed system, which is actually the most simple, we shall generally be able to deduce from the primitive equations those which correspond to them in this special construction, by transforming them, by two successive eliminations, into two equations, each containing only two of the variable co-ordinates, and thereby corresponding to the two surfaces of projection. Such is really the principal destination of this sort of geometrical combination, which thus offers to us an invariable and certain means of recognizing the identity of lines in spite of the diversity of their equations, which is sometimes very great.

IMPERFECTIONS OF ANALYTICAL GEOMETRY.

Having now considered the fundamental conception of analytical geometry under its principal elementary aspects, it is proper, in order to make the sketch complete, to notice here the general imperfections yet presented by this conception with respect to both geometry and to analysis.

Relatively to geometry, we must remark that the equations are as yet adapted to represent only entire geometrical loci, and not at all determinate portions of those loci. It would, however, be necessary, in some circumstances, to be able to express analytically a part of a line or of a surface, or even a *discontinuous* line or surface, composed of a series of sections belonging to distinct geometrical figures, such as the contour of a polygon, or the surface of a polyhedron. Thermology, especially, often gives rise to such considerations, to which our present analytical geometry is necessarily inapplicable. The labours of M. Fourier on discontinuous functions have, however, begun to fill up this great gap, and have thereby introduced a new and essential improvement into the fundamental conception of Descartes. But this manner of representing heterogeneous or partial figures, being founded on the employment of trigonometrical series proceeding according to the sines of an infinite series of multiple arcs, or on the use of certain definite integrals equivalent to those series, and the general integral of which is unknown, presents as yet too much complication to admit of being immediately introduced into the system of analytical geometry.

Relatively to analysis, we must begin by observing

that our inability to conceive a geometrical representation of equations containing four, five, or more variables, analogous to those representations which all equations of two or of three variables admit, must not be viewed as an imperfection of our system of analytical geometry, for it evidently belongs to the very nature of the subject. Analysis being necessarily more general than geometry, since it relates to all possible phenomena, it would be very unphilosophical to desire always to find among geometrical phenomena alone a concrete representation of all the laws which analysis can express.

There exists, however, another imperfection of less importance, which must really be viewed as proceeding from the manner in which we conceive analytical geometry. It consists in the evident incompleteness of our present representation of equations of two or of three variables by lines or surfaces, inasmuch as in the construction of the geometric locus we pay regard only to the *real* solutions of equations, without at all noticing any *imaginary* solutions. The general course of these last should, however, by its nature, be quite as susceptible as that of the others of a geometrical representation. It follows from this omission that the graphic picture of the equation is constantly imperfect, and sometimes even so much so that there is no geometric representation at all when the equation admits of only imaginary solutions. But, even in this last case, we evidently ought to be able to distinguish between equations as different in themselves as these, for example,

$$x^2+y^2+1=0,\ x^6+y^4+1=0,\ y^2+e^x=0.$$

We know, moreover, that this principal imperfection often brings with it, in analytical geometry of two or of three dimensions, a number of secondary inconveniences, arising from several analytical modifications not corresponding to any geometrical phenomena.

Our philosophical exposition of the fundamental conception of analytical geometry shows us clearly that this science consists essentially in determining what is the general analytical expression of such or such a geometrical phenomenon belonging to lines or to surfaces; and, reciprocally, in discovering the geometrical interpretation of such or such an analytical consideration. A detailed examination of the most important general questions would show us how geometers have succeeded in actually establishing this beautiful harmony, and in thus imprinting on geometrical science, regarded as a whole, its present eminently perfect character of rationality and of simplicity.

Note.—The author devotes the two following chapters of his course to the more detailed examination of Analytical Geometry of two and of three dimensions; but his subsequent publication of a separate work upon this branch of mathematics has been thought to render unnecessary the reproduction of these two chapters in the present volume.

THE END.

DOVER PHOENIX EDITIONS

An Introduction to the Approximation of Functions, Theodore J. Rivlin. Unabridged republication of the 1969 edition. 160pp. 49554-X

An Essay on the Foundations of Geometry, Bertrand Russell. Unabridged republication of the 1897 edition. 224pp. 49555-8

Elements of Number Theory, I. M. Vinogradov. Unabridged republication of the first edition, 1954. 240pp. 49530-2

Asymptotic Expansions for Ordinary Differential Equations, Wolfgang Wasow. Unabridged republication of the 1976 corrected, slightly enlarged reprint of the original 1965 edition. 384pp. 49518-3

Physics

Semiconductor Statistics, J. S. Blakemore. Unabridged, corrected, and slightly enlarged republication of the 1962 edition. 141 illustrations. xviii+318pp. 49502-7

Wave Propagation in Periodic Structures, L. Brillouin. Unabridged republication of the 1946 edition. 131 illustrations. 272pp. 49556-6

The Conceptual Foundations of the Statistical Approach in Mechanics, Paul and Tatiana Ehrenfest. Unabridged republication of the 1959 edition. 128pp. 49504-3

The Analytical Theory of Heat, Joseph Fourier. Unabridged republication of the 1878 edition. 20 figures. 496pp. 49531-0

States of Matter, David L. Goodstein. Unabridged republication of the 1975 edition. 154 figures. 4 tables. 512pp. 49506-X

The Principles of Mechanics, Heinrich Hertz. Unabridged republication of the 1900 edition. 320pp. 49557-4

Thermodynamics of Small Systems, Terrell L. Hill. Unabridged and corrected republication in one volume of the two-volume edition published in 1963–1964. 32 illustrations. 408pp. 6½ x 9¼. 49509-4

Theoretical Physics, A. S. Kompaneyets. Unabridged republication of the 1961 edition. 56 figures. 592pp. 49532-9

Quantum Mechanics, H. A. Kramers. Unabridged republication of the 1957 edition. 14 figures. 512pp. 49533-7

The Theory of Electrons, H. A. Lorentz. Unabridged reproduction of the 1915 edition. 9 figures. 352pp. 49558-2

The Principles of Physical Optics, Ernst Mach. Unabridged republication of the 1926 edition. 279 figures. 10 portraits. 336pp. 49559-0

The Scientific Papers of James Clerk Maxwell, James Clerk Maxwell. Unabridged republication of the 1890 edition. 197 figures. 39 tables. Total of 1,456pp.
Volume I (640pp.) 49560-4; *Volume II* (816pp.) 49561-2

Vectors and Tensors in Crystallography, Donald E. Sands. Unabridged and corrected republication of the 1982 edition. xviii+228pp. 49516-7

Principles of Mechanics and Dynamics, Sir William (Lord Kelvin) Thompson and Peter Guthrie Tait. Unabridged republication of the 1912 edition. 168 diagrams. Total of 1,088pp. *Volume I* (528pp.) 49562-0; *Volume II* (560pp.) 49563-9

Treatise on Irreversible and Statistical Thermophysics: An Introduction to Nonclassical Thermodynamics, Wolfgang Yourgrau, Alwyn van der Merwe, and Gough Raw. Unabridged, corrected republication of the 1966 edition. xx+268pp. 49519-1

Engineering

Principles of Aeroelasticity, Raymond L. Bisplinghoff and Holt Ashley. Unabridged, corrected republication of the original 1962 edition. xi+527pp. 49500-0

Statics of Deformable Solids, Raymond L. Bisplinghoff, James W. Mar, and Theodore H. H. Pian. Unabridged and corrected Dover republication of the edition published in 1965. 376 illustrations. xii+322pp. 6½ x 9¼. 49501-9